PHARMACEUTICAL PRODUCTION FACILITIES

Design and Applications

Second Edition

Graham C. Cole

Pharmaceutical Process Specialist
Higham, Suffolk

CRC Press
Taylor & Francis Group
Boca Raton London New York

CRC Press is an imprint of the
Taylor & Francis Group, an **informa** business

CRC Press
Taylor & Francis Group
6000 Broken Sound Parkway NW, Suite 300
Boca Raton, FL 33487-2742

First issued in paperback 2019

© 2010 by Taylor & Francis Group, LLC
CRC Press is an imprint of Taylor & Francis Group, an Informa business

No claim to original U.S. Government works

ISBN-13: 978-0-367-40062-0

Visit the Taylor & Francis Web site at
http://www.taylorandfrancis.com

and the CRC Press Web site at
http://www.crcpress.com

PHARMACEUTICAL PRODUCTION FACILITIES
Design and Applications

Contents

Contributors

Nigel Stephen Cole has an MSc from Bradford University in Chemical Engineering and has spent 10 years in the process automation industry. Six years were spent with a leading supplier of process control systems and four years as an information technology project manager with GlaxoWellcome. At GlaxoWellcome he has been implementing manufacturing execution systems and shop floor solutions.

Michael Walden, who provided the excellent material for Chapter 7 and a new section in Chapter 6, is registered as an architect in the United Kingdom, and has worked on projects in the pharmaceutical industry virtually exclusively for the past 20 years.

His experience includes a period working as project manager with the International Facilities Group of American healthcare company, Baxter Travenol and time with several major design build contractors. Following a period as Technical Director of a cleanroom construction company Michael Walden has for the last six years worked with Clean Design involved with projects in the United Kingdom, Europe and South East Asia.

Michael holds a diploma in Architecture from Leeds and is a corporate member of the Royal Institute of British Architects and a Member of the Society of Environmental Engineers.

Preface

Thus grew the tale of Wonderland,
Thus slowly, one by one,
Its quaint events were hammered out
And now the tale is done,
And home we steer, a merry crew,
Beneath the setting sun.

Lewis Carroll: *Alice's Adventures in Wonderland*

The regulatory requirements for production facilities within the pharmaceutical industry are being continually and incessantly tightened. On the one hand this leads to a higher quality product on a consistent basis, but on the other it has made production itself an increasingly costly part of the overall process of drug discovery, development, manufacture and distribution.

In 1990, in the first edition of this book, Graham Cole made an excellent attempt to set out the requirements of the major elements of the facilities required for secondary pharmaceutical manufacture (i.e. the manufacture of the dosage form rather than the bulk drug chemical). This was the first book of its type to attempt this and it made a valuable contribution. However, due to the current ever changing practice it was inevitable that a second edition would become necessary. Graham Cole has therefore produced this new, up-to-date version of the text.

The first edition has been substantially enlarged and revised. In this second edition, Graham retains the basic structure of the previous edition but has considerably updated and modified the text, particularly in the areas of process flow, architectural considerations, clean room design (including a discussion of the latest regulations), tablet coating systems, special production systems and, inevitably, validation. This revised edition serves its purpose well – it is an excellent introduction to concepts and practical problems associated with pharmaceutical secondary manufacturing.

The book illustrates the many concepts and constraints and 'state of the art' control that must be considered in the design of small-, medium- and large-scale production plants. The layout of a facility and the flow of materials and personnel through the facility are considered with reference to ensuring compliance with current good manufacturing practice.

The benefits that can be obtained from using totally automated enclosed systems for small production runs are demonstrated using sterile operations and the manufacture of solid dosage forms as examples. Mike Walden shows how clean rooms have developed and how the latest regulations affect their design. The latest techniques for reducing contamination levels from the operator and the product are discussed. A solid dosage facility is used as a model to show how integration can be achieved with a corresponding minimization of costs and improvement in productivity, and even in product elegance. The ideal facilities are outlined, illustrating the design considerations that need to be applied in modern dosage manufacturing facilities in order to enable the systems to be validated to current standards.

The chapters on validation have been updated to include current thinking on process and computer control, such as the *Guide to Automated Manufacturing Practice* document (GAMP). New concepts such as Computer Integrated Manufacturing (CIM) and Just in Time (JIT) manufacturing are discussed.

Secondary production is no longer an add-on to the overall process of medicines design and manufacture. In the early years the pharmaceutical industry did not invest heavily in dosage form production facilities because the cost of the drug itself was so high that manufacturing costs were a minor component of the total cost of medicines and any inefficiencies in that area were of no great importance. Now production has to be much more efficient as medicines costs are driven downwards by governments and by the ever increasing pressure from generic products. The industry has had to become more sophisticated and its production procedures increasingly efficient. The extensive manual procedures of yesteryear have virtually disappeared and process automation has improved out of all recognition. Most of the recent pharmaceutical company mergers have been driven by the design for increased efficiency of all aspects of drug discovery – including rationalizing production facilities on a worldwide basis.

It is also inevitable that the rate of changes that we have seen over the last few years will never plateau. We will be forced forever into greater and greater sophistication – driven either by financial or regulatory requirements. So we thank Graham for his efforts in collating this valuable and timely information into one text. This single book will not tell you all that you need to know about pharmaceutical production facilities, but it is the ideal starting point to enable you to understand the concepts which lie behind the problems before you have to delve into a detailed consideration of the regulations themselves.

Michael Aulton
Professor of Pharmaceutical Technology and
Head of Pharmaceutics
De Montford University
Leicester, UK

1

Introduction

'When I use a word', Humpty Dumpty said, in rather a scornful tone, 'it means just what I choose it to mean – neither more nor less'.

The pharmaceutical industry has undergone fundamental changes and restructuring since the Second World War. The substantial and rapid progress that has been made is based on research and development of new compounds with outstanding rewards for those companies that have developed 'winners'. Products like ALDOMET, TAGAMET, and ZANTAC have rewarded those companies with sales running into thousands of millions of dollars. The costs in research and development alone (safety testing programmes, clinical trials, manufacture and marketing costs) to market one of these products is probably in the region of between 150–200 million US dollars. One result of these costs has been the mergers and restructuring that have taken place between companies: Glaxo and The Wellcome Trust; SmithKline Beckman and Beecham; Squibb and Bristol-Myers; Rhone Poulenc-Rorer; Roussel and Hoechst; Astra and Fisons; and Boots Pharmaceutical Division and BASF, to name but a few. These will continue as the top companies strive to maintain their market share and the Japanese attempt to increase their presence outside of Japan in the world market. These mergers have been designed to minimize spiralling costs in research and development, rationalize the research base by concentrating on specific therapeutic classes, and maximizing productivity in drugs manufacture.

The rewards in the pharmaceutical industry have been the result of developing unique chemotherapeutic products, and traditionally these products were manufactured by using a collection of different types of facility, a variety of construction materials, and equipment generally borrowed from other industries. For example the planetary mixer originated in the bakery. Probably the tablet machine with systems to regulate its operation is the best example of innovation in the industry. Computer integrated manufacture has taken a lot longer to develop for plants designed to produce sophisticated modern dosage forms.

However, with the impact of regulatory authorities such as the Medicines Control Agency and the US Food and Drug Administration, and the rationalization of manufacturing to supply various markets, this position is changing. There is no manual or textbook which details all the design requirements for a modern pharmaceutical plant; there is, however, a collection of documents such as *The Orange Guide*, *The Code of Federal Regulations*, and various guidelines whose impact necessitates that task forces formed within manufacturing companies require a much higher and broader level of ability and skills. Computer controlled processes are becoming the norm with plants like the Merck facilities in Europe and the USA, the Pharmachemie Plant in Holland, and many others.

There should be a comprehensive system so designed, documented, implemented and controlled and so furnished with personnel, equipment, and other resources as to provide assurance that products will be consistently of a quality appropriate to their intended use. The attainment of this quality objective requires involvement and commitment of all concerned at all stages (*The Orange Guide* 1983).

The pharmaceutical industry is unique in the procedures and methods of manufacture that it uses to ensure the integrity of the products it produces. These are essentially achieved by three main functions: current Good Manufacturing Practice (cGMP), Quality Assurance (QA), and Quality Control (QC). This book will be concerned with the design and operation of secondary manufacturing facilities and the involvement of these three main functional areas will be assessed in terms of the overall strategy of operating a plant of this type. The manufacturing side of the industry can be divided into two parts, and the basis that will be used here considers primary manufacture as the production of the bulk drug as a fine chemical, using a variety of routes such as organic synthetic chemistry, fermentation, or biotechnology. Secondary manufacture manipulates the drug form, using various excipients to produce a packaged dosage form.

The word 'design' can be interpreted to mean different things to different people. For the purpose of this book it will be defined as the preparation of drawing a plan or preliminary sketch in the process of using an invention, and it follows then that certain essential criteria must be fed into the design to ensure that it can be confirmed. It would be expected that with the volume of papers and articles currently published on the design of buildings specifically for the pharmaceutical industry, the use of sophisticated modelling of expert systems, optimization of process systems, and project engineering concepts that it would be possible to conceptually design and build a secondary manufacturing pharmaceutical facility without great difficulty. This is still true as long as the project remains simple and all parts could be visualized; potential problems could be analysed rapidly and an alternative solution found.

In the past most new facilities were designed as a building into which items of equipment were positioned after its completion. Manufacturing operations were considered to be a series of unit operations that could be arranged in almost any format. Dispensing, powder blending, granulation, tablet compressing, and tablet coating were generally looked upon as separate operations and could be located in almost any part of the building. There was no integration of the manufacturing system. This led to extensive development of equipment to provide self-contained operations – 'Islands of Automation'. For example table presses were produced with control and monitoring systems which reduced the operator's role and enabled the operator to run more than one item of equipment.

However, a number of factors combined to concentrate companies' corporate minds on the manufacturing operations (the poor but essential relation). A comparison between the amount of money spent on research blocks, offices, and marketing facilities shows where the companies believe their priorities lie. However, without comparable investment in production, designed to prevent product mixup, the recall of a product from the market can easily cost millions of dollars.

These factors can be divided into two classes, those associated with the regulatory authorities ('what we are compelled to do we will') and the technology push development. Fluid bed drying replaced tray drying with dramatic improvements in productivity. The replacement of starch paste using pre-gelatinized starch and cold water, simplified the development of automated granulation equipment, such as the high speed high shear mixer/granulator. Microwave drying appears to be adding a further dimension to this technique.

The advent of the oil crisis forced many companies to examine their manufacturing facility design policy to ensure a more efficient use of resources. The formation of the Common Market required rationalization of products manufacture, and 1992 provided more stimulus to the location of plants for the manufacture of single products, or groups of similar products. This has created the opportunity to design systems that are more efficient and provide the economy of scale. The use of automated handling systems, automated guided vehicles (AGVs) and automated warehouses have become the realities of life and not just ideas. The added advantage of rationalization is the allocation of valuable raw materials, strategic planning, and the development of worldwide inventory control.

New products that are being developed by research based companies have high potency, low volume, and, together with the sustained release preparations which demand higher levels of technology in their manufacture, require smaller but more complex processes.

However, facilities are now becoming increasingly complex; what was once a labour intensive industry is becoming more capital intensive as companies build new facilities using computer controlled manufacturing processes, and large multinational companies are planning and controlling their raw materials and products on a worldwide basis.

Four main areas of design will be considered:

• process
• facility layout and design
• site selection
• the impact of environmental constraints.

The process will be examined from the reception of the raw material (chemical or packaging component) through to the warehousing of the packed product.

In the past the manufacturing facility was very localized. Management organized the flow of materials and products to a number of markets. They were concerned only with the profitability of their plant and were largely insulated from problems in other parts of the world. If they were in profit then the parent company tended to 'leave well alone'. The infrastructure of the subsidiary was not geared to develop a worldwide strategy in terms of manufacturing philosophy and control of raw materials.

All the largest multinationals operate a central engineering group which has overall responsibility for facilities both at home and overseas, and associated with this group is a team responsible for the process and manufacturing operation.

Most of these centralized departments have attempted to impose:

• standardization of formulations
• standardization of manufacturing equipment.

3

This is a good theory and has had a limited amount of success. It collapses where countries have limited amounts of foreign exchange, impose import controls, generally impose limitations and conditions on the building of new facilities, introduction of new products into their market, transfer of profits outside the country, and the use of local raw materials. Compromises have to be made where there is a policy implementing a worldwide formulation for manufacture requiring the same equipment and similar environmental conditions.

The industry also suffers from the problems of small volumes (batches) using expensive ingredients and in latter years the attention of regulatory authorities like the MCA, the European Medicines Evaluation Agency (EMEA), and the FDA. The development of equipment has far outstripped the system used to integrate this equipment into continuous batch operation. Merck in the early 1970s developed an automated dedicated batch process which handled the raw material from the point at which it entered the system and ended where the dosage form was ready for packaging.

Although the material handling concept was partly the same, SmithKline & French (SKF), now SmithKline Beecham (SKB) developed along a different route. It did not have the high volume of dosages that Merck used to justify its dedicated plants (i.e. 1000 M/annum/single product). The SKF plants at Alcala in Spain and Milan in Italy used what has become known as the Lhoest Principle in an attempt to overcome the disadvantages of the dedicated plant and provide for the small number of dosages manufactured/annum and a large number of products. This added a large degree of flexibility to the operation. However, the Milan plant was manufacturing fewer than 20 products and one of these accounted for greater than 75% of the total production. In both types of plant the objective was to use totally enclosed systems for transfer of materials and product and standardization of the containers. In solid dosage manufacture dust has always been the main problem in attempting to eliminate cross-contamination especially in the transfer operation. The Pharmachemie Plant in Holland was developed on a similar mode but handles between 150 and 200 products.

What is important to recognize is that the pharmaceutical industry has some special requirements that need to be interpreted correctly. To do this requires an understanding of the jargon. Some examples are:

- (cGMP) current Good Manufacturing Practices
- Validation
- FDA (Food & Drug Administration)
- IDIP (Intensified Drug Inspection Programme)

which fall so lightly from those professionals employed in the industry.

This book is an attempt to harness some of these ideas together with concepts such as Validation to help provide a route for both scientists and engineers through these minefields.

My thanks are due to my wife for her indulgence in providing the secretarial expertise and my colleagues both past and present to whom I owe a debt of gratitude.

We should also not forget Lewis Carroll whose lines in *Alice's Adventures in Wonderland* and *Through the Looking-Glass and What Alice Found There* are still as appropriate today as they were in 1865 when first published.

> *'Let the jury consider their verdict' the King said, for about the twentieth time that day.*
> *'No. No' said the Queen, 'Sentence first – verdict afterwards'.*

2

Project Design and Management

'I should see the garden far better', said Alice to herself, *'if I could get to the top of that hill: and here's a path that leads straight to it – at least, no, it doesn't do that –'*

Design includes all the aspects of process engineering, building, plant design, environmental services and validation of any new facility, modification or refurbishment; and there are typically a number of attributes and sequential stages of a project that have to be studied.

- Feasibility
- Process engineering
  - □ Personnel flow
  - □ Process utilities
- Material flow
  - □ Layout
- Construction and validation.

During the feasibility study, the attributes that are critical to the success of the project and its operation mean that it has to be viable in terms of being practicable. It must be able to be operated satisfactorily in terms of producing a consistent product within the defined specifications and produce the required quantities and range of products. Its reliability must be such that it can be operated during its expected life cycle within the defined operating conditions and costs.

All projects must be constrained within strict financial limits based on the return on investment (ROI), and an expenditure ceiling must be applied on the operating costs of the plant. If any of these are allowed to slip then the company can be disadvantaged when selling in a competitive market place.

An important aspect of the project to consider at this stage is under what regulations will the plant or new facility have to operate. In particular does the MCA, EMEA, or FDA have a role to play in the design? Does the plant require validation (well thought out, well structured, well documented common sense)? Validation has tended to concentrate on secondary manufacturing systems, but it is now being increasingly extended to primary manufacturing facilities.

Feasibility

The life cycle of any project follows a regular pattern, and what is feasible and what is not must be closely examined. All instigators of projects and those who become assigned to the project team tend to look at what is available in the market place and what technically is available, and this generally produces a concept that far exceeds the financial resources of the company.

There are many commercial pressures to proceed to contract at the earliest possible date, but the utmost care is needed at this point. Up front decisions have a major impact on the long term viability of a project, and the design stage should be fully worked out before placing contracts.

In comparison, the detailed design is unlikely to affect the overall project viability, but will contribute to the ease or difficulty in commissioning, validating, operating, and maintaining the plant.

A production plant has many separate parts, and it is impossible to set about putting those parts together without a comprehensive project definition or brief. Most corporations will grant approval for a project on outline information, and so the project manager must develop the brief with his project team.

Market Requirements

What is the product definition?

How many units are required per shift/day/week/year?
What are the range of product sizes/contents/material?
What are the foreseen future market trends?

A spreadsheet needs to be generated to define these essentials.

Regulatory Requirements

Which regulatory authorities will approve the product and inspect the operations?
What are the anticipated standards required for the design, validation, and documentation of the process and its operative environment?

Financial Constraints

What is return on investment (ROI) expectation?
What are the capital constraints/limits?
What is the foreseen market price for the product/profit margins/maximum acceptable cost of production (COP)?

Timing

When is the product expected on the market and at what volume?
Define a realistic time frame for the project (neither pessimistic nor optimistic) and rigidly plan and control the project against this schedule.

Conceptual Location

Each project is likely to have had at least a conceptual location when submitted for management approval (i.e. a specific green field site, or space in an existing operational area). The location should be challenged on an ongoing basis, until the contracts are let – there is often a more suitable location.

Process Constraints

Establish a list of constraints or special process needs which will affect the design or choice of location. Typical examples would be as follows:

What hazardous materials will be used (toxic, inflammable, explosive)?
Are there effluent problems (airborne, drainage, special disposal)?
What equipment will be required (unusual services, vibration-free, access for installation, maintenance, etc.)?
Are automated systems to be considered, and what transfer systems will be used?

It is necessary at this stage to consider the synthesis by which all parts of the project are put together.

Macro Process Review

The first stage of the synthesis is to review the project on a macro scale, starting with the manufacturing process, followed by ancillary plant and services, support areas, and storage needs.

Manufacturing Process

What type of process – continuous or batch?
What type of equipment – existing, or special that requires development? What is the output per item of equipment, what space is required, based on equipment size and product flow configuration, what are the services demands, and what are the special environmental needs for the process (and regulatory constraints)?

Personnel Flow

What are the personnel requirements for this operation? Do we know? Companies are becoming more concerned and are restricting the number of operatives in the plant. 'Head counts' are a major consideration, and to maximize their efficient use, the layout of equipment, changing facilities, and offices must be carefully studied.

Process Utilities

What type and size of ancillary plant is needed to support the process and provide the necessary services and utilities? What are the specifications?

HVAC – quality of air/Class 10 000
Water – city/treated/purified/water for injection (WFI)
Steam – clean
Compressed air – grades required, breathing, etc.
Electricity
Vacuum
Special gases
Effluent
Inert gases

Support Areas

What type of support areas are necessary to facilitate viable operation of the process and to provide for the needs of personnel? Included in this category would be product and materials testing facilities, engineering and maintenance areas, locker and change rooms, offices, cafeteria and personnel needs, goods receiving areas and warehousing.

Material Flow

These needs must be identified with reasonable accuracy at an early stage, since these are the most likely to determine the viability of a particular location, and the list would include, at the very least, the following:

Raw materials:	Reception/quarantine area (requirement)
	Released material area
Manufacturing	Subassemblies
	In process materials
	In process bulks
	Part processed materials
	Products awaiting packaging
	Packaging
Finished goods	Quarantine area (requirement)
	Released material area
Rejected materials	Raw materials
	Finished goods

At this stage, the project team will have available to them a list of space requirements and sketches of individual needs for each of those areas. For the key project areas, it is likely that design criteria sheets would also have been prepared. These are a useful tool for circulating to various functional groups for sign-off, before the detailed engineering design process commences.

Plant Layout

Armed with the information from the macro process review, it is now possible to prepare an initial plant or process area layout.

Flow patterns must be logical and simple, and route lengths kept to a minimum. Materials routes and locations should make easy the segregation of work in progress, sterile and non-sterile product, quarantined and released materials. Personnel and materials routes should be kept apart. Utilities should be directly routed from plant rooms to process areas and should be grouped together whenever possible.

The space allocated to each area of the process should be in line with previously defined needs, with agreed allowance for growth. Clean areas should be kept to a minimum volume, in view of the high unit cost of such areas.

The layout should facilitate future expansion of the process – avoid critical processes installed against outside walls or sandwiched between other critical processes.

Budget Costs

Budget costs can be prepared for the project as foreseen:

Project capital costs	– Equipment and installation costs
	– Facility construction
	– Utilities equipment and their installation
	– Validation
Future operating costs	– Labour
	– Materials
	– Services
	– Other overheads

Programme

Enough information is now at hand to enable a realistic construction, equipment installation, and commissioning programme to be prepared for the project. An accurate assessment of the start-up learning curve should also be possible.

A typical programme to be inserted here is shown in Figure 2.1, and the costs of the project stages are illustrated in Figure 2.2.

Project Assessment

The project must be re-assessed at this stage, before the detailed design is undertaken, or any commitments are made. Regular reviews of this kind should be undertaken at reasonable intervals throughout the execution phase. These reviews should follow the loop shown in Figure 2.3.

The detailed design of the plant will now have taken on board the following aspects:

- *Process routes*

Well defined process flow diagrams which include all existing products and any that may be in the research and development pipeline.

- *Equipment layout*

Layout of equipment and ancillary plant areas must be developed as soon as details of the individual items of equipment are known. Dimensions of the areas must

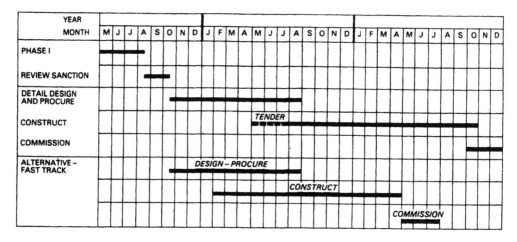

Figure 2.1 Typical overall project schedule for a pharmaceutical production unit

permit equipment to be installed to operate and to be maintained and space must be adequate for persons performing those functions.

The material flow within the space, and its transfer to the next stage in the operation, must be satisfactory. Common changing areas should be used wherever possible.

Overall, the use of floor and volume within the buildings must be efficient, as construction costs are high. Some examples are (1997):

Building shells are typically	£450–550 per m^2
Fitting out costs non-sterile/aseptic	£1200–1500 per m^2
Fitting out costs sterile	£2000 + per m^2
Estate costs	£100–150 per m^2

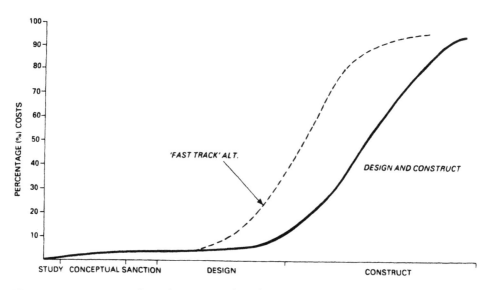

Figure 2.2 Project stages for a pharmaceutical production unit

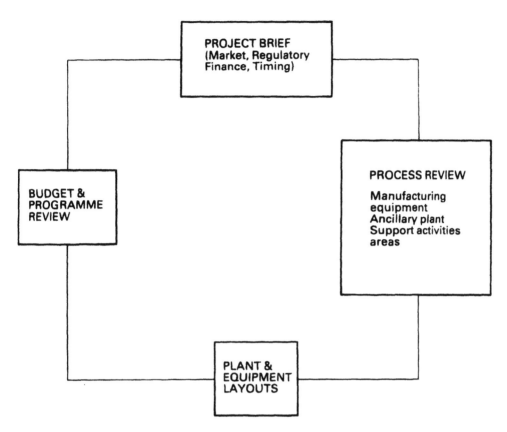

Figure 2.3 Project reviews

Finally, the impact of layout on the services design must be considered, as unnecessary lengths of ducting, pipework, and cable route all add to cost and the length of the construction programme.

● *Construction details and schedule of finishes*

Room data sheets are good examples. Most companies have defined facility construction and finishes standards which have been established over a period. These would specify, by the category of area, acceptable types of finishes and the design details. New materials and details should be evaluated with care before major installations are undertaken.

● *Floors*

These are the most critical of all finishes and the most often wrongly selected. Choice ranges from flat floors to those laid to gradient for drainage. Typical finishes are:

Power float finish concrete
Paint finish over concrete
Vinyl (tile, sheet, welded)
Self-levelling screeds (epoxy, polyurethane, etc.)
Epoxy aggregate screeds.

● *Walls*

Construction may be from a number of materials, block, brick, plasterboard, demountable partitioning, glass, etc., with finishes from emulsion paint over plaster, epoxy paint, stove enamelled steel through GRP panels, spray-on PVC/PVA, to welded vinyl sheet.

● *Ceilings*

Simple ceilings may be painted over structural concrete, painted or vinyl covered gypsum plaster. Lay-in ceilings may use vinyl covered tiles, GRP panels, or stove enamelled steel panels.

● *Details*

The success of design depends on attention to detail:

☐ the design of doors and windows
☐ services penetrations
☐ special fittings, i.e. pair locks
☐ corners and joints
☐ ventilation duct.

● *Equipment details*

As well as functioning correctly, the detailed design of equipment must extend to materials and the finishes specification. It must always be readily cleanable, crevice-free in all areas, and in a finish compatible with the process and product. Construction details and standards must be defined in order to achieve the following:

☐ Keep particulate counts low (non-dusting or shedding but easily cleanable).
☐ Keep microorganism counts low (easily cleanable, impervious, and non-mould supporting).
☐ Be chemical resistant (non-degrading after sustained contact with product, process chemicals, or cleaning agents).
☐ Be physically durable (resistant to steam, moisture, and the actions of idiots!).
☐ Be easily maintained.

● *Environmental air*

The fundamental qualities required of an environmental air system are:

☐ Quality (particles, size per m^3)
☐ Quantity (volume flow)
☐ Flow patterns
☐ Pressurization of enclosed space
☐ Temperature and humidity.

Each is interlinked with the others in affecting conditions. Various design codes reflect the experienced minima necessary to achieve certain environmental standards, but processes can demand a facility in excess of these minimums.

□ Air quantity (air changes per hour)

The correct quantity is necessary to displace particles from the environment, pressurize the space, and control the temperature and humidity.

Changes per hour	Class
>120	100
>40	10 000
>20	300 000

(Federal standard 209E) (BS 5295)

□ Air quality

The quality, location and standard of maintenance of filters affects the air quality incoming to a space. Selection depends on the process burden and size range of particulates, but would generally be as follows:

In AHU panel filters	Unclassified environment
In AHU bag filters	Class 300 000
Terminal HEPA filters	Class 10 000

□ Air flow patterns

General recommendations can be found in the appropriate international standards, but a symmetrical supply and exhaust configuration is needed. Low-level exhausts are likely to be needed for Class 100 000 rooms or better.

□ Pressurization

Code requirements are for rooms to be pressurized in discrete steps from cleanest rooms down to the surrounding base. These steps are sometimes impracticable when process demands are superimposed.

(1) Containment of process materials (i.e. drugs) is contrary to room pressurization.
(2) Air leakage through apertures gives excessive volume losses or affects product.

● *Temperature and humidity*

Clean rooms can be very uncomfortable for heavily gowned production staff unless temperature and humidity are closely controlled. Process equipment gives off much heat, and product quality can be badly affected by severe temperatures or humidity. Plant size and air change rates to control these may exceed other environmental requirements.

The overall ventilation system for a set of process areas must be designed to ensure compatibility between adjacent areas, permit the use of common plant, and ensure future balances are acceptable.

● *Water services*

Water services must be designed, starting with a plant overview identifying qualities of water for each part of the process, and daily and peak usage rates; and they must be based on the quality of the incoming municipal or well water. Pipe, storage tank, and treatment plant sizes must be calculated for the planned duty and for future growth. Oversized plant can give as many problems as undersized!

A plant schematic would typically be prepared at this stage and would then lead to the sizing and layout of water treatment plant areas and distribution pipework.

All water service installations must be carefully designed, installed, and documented, to ensure consistent and acceptable water quality at each stage of the process. This, in particular, applies to high purity water systems, where stringent installation and operational controls are critical.

- *Steam*

Steam services are critical and are one of the most expensive utilities in a sterile products plant. Principal uses are for sterilization, water for injection production, and some process heating functions. Plant sizing is important to ensure that consistent steam supplies are available.

Economic plant design dictates that the boilerhouse should be at the centre of gravity of steam usage, normally adjacent to sterilization. Steam and condense services are difficult to maintain in a cGMP condition and should be kept to a minimum in production areas.

- *Compressed air*

Compressed air is the most often abused service in facilities, and sterile products plants are not necessarily exceptions to this.

The compressed air system must be well designed from compressors through to points of use:

□ Well engineered compressor room with back-up.

□ All pipework correctly sized, to slopes and drainable.

□ Simple distribution pipework, well laid out with top-off connections.

□ Complete system cleaned, tested, and documented.

Categories of compressed air may be as follows:

Process air – oil-free, particle-free, dry, and suitable after terminal filtration for product contact.

Plant air – particle-free, dry, and suitable for pneumatic equipment.

Breathing air – as process air but hepafiltered.

Instrument air – as breathing air.

Process air may be used as plant air, so long as back-flow prevention equipment is installed upstream of oilers. Exhaust air from equipment, if contaminated, must be correctly handled to avoid environmental contamination.

- *Electrical services*

Adequate and safe electrical services are as important a part of a plant requirement as any other process plant – process power, small power, and lighting. Special consideration must be given to the following:

□ Equipment connections: Are individual cables, conduits, or trunking acceptable, or are service drops necessary? Is equipment to be hard-wired, or are socket disconnects needed? What are the special clean area or wet environment requirements?

□ Control panels: Location of control panels must be defined for functionability, ease of installation, and maintenance. Location dictates construction standard.

□ Lighting: Lighting levels must be adequate for the process, and fittings must suit the environment:

(1) Intrinsically safe
(2) Waterproof/multi-gasketted
(3) Sealed.

What areas of the plant require flameproof systems?

● *Vacuum and special gases*

Vacuum pumps are either dry (oil ring pumps) or wet (water sealed pumps or steam jets). Wet processes such as rinsing or product evacuation must use a wet pump, and in view of the high risk of contamination, must be capable of being sanitized.

Vacuum pumps must be located as close as possible to the process, for efficient vacuum and pipes must be drainable and capable of being decontaminated.

Special gases are becoming an increasing requirement these days. A location close to the use point must be selected for cylinder or bulk supplies of nitrogen, helium, etc., and suitable engineering standards must be defined for the installation. Usually, compressed air standards are accepted as a minimum.

● *Effluent*

Last, but not least of the provision, must be effluent handling systems. The 'out of sight, out of mind' attitude can cause major problems, whereas these systems need to be well thought out and installed at an early stage of a project.

Drains: These receive effluent at a range of temperatures with a variety of chemicals, pollutants, and other undesirables. They are mostly inaccessible for repair and replacement and are a potential cause of contamination of the working environment. Stainless steel drains and manholes for all process drainage systems within buildings are now advocated. Drains must be adequately sized, having sufficient slope, and be fitted with back flow preventers, where process, surface, and foul drainage interconnect. No critical process system should ever be directly connected into a drainage system.

Effluent handling systems: These must protect the environment from the process.

□ Lagoons and separator stations for process and domestic wastes.
□ Trickle tanks to avoid problems with peak discharge conditions.
□ Neutralization tanks for the rectification of chemical imbalance.
□ Holding tanks for emergency discharges or materials to be disposed of by other means.

These provisions must be made at the very start of the project, and must be correct.

OVERALL DESIGN CONCEPT

For the development of new projects the conceptual ideas must be brainstormed to provide a basis for design and to evaluate all possible scenarios. An example of the execution plan and project stages is given here for the development of conceptual design.

Execution Plan Project Stages

The execution of the work required to produce the conceptual design falls into four stages:

(1) Client brief evaluation and assessment.
(2) Process development.
(3) Conceptual design based on the client brief, outcome of any previous studies, and finalized layout.
(4) Costing.

Depending on the complexity of the plant, these four stages can be divided up, allowing a total period of 13 weeks to complete the work.

Stage 1 – Client brief evaluation, Weeks 1 and 2
This entails a complete evaluation of the client's brief to establish the basis of the conceptual design. It will include discussions between the client and the contractors' or consultants' key team members, for example the process engineer and architect.

Stage 2 – Process development, Weeks 3 to 6
Having fully assessed the client's ideas and requirements the process engineer and, if necessary, a specialist consultant, will develop the process design.

They will study alternative solutions, optimization of batch sizes, selection of key equipment items, and development of the overall layout.

During this period assistance will be provided on a part-time basis by other specialist groups, i.e. pharmaceutical architects, heating, ventilation and air conditioning engineers, materials handling and instrumentation and control engineers.

During week 6 the proposed process and sketched layout are reviewed in depth with the client. Alternatives are then considered.

On completion of the review the favoured process solution will have been determined.

Stage 3 – Conceptual design, Weeks 7 to 10
At the commencement of week 7 the full team commences full development of the conceptual design. The scope of work to be carried out and the deliverable items are defined at this stage.

By the end of week 10 the concept is fully defined in drawings, specifications, and written documents. Further reviews with the client take place before estimating the cost.

Stage 4 – Cost estimation and document preparation, Weeks 11 to 13
The final three weeks of the study comprises:

☐ Preparation of cost estimate
☐ Finalization of conceptual design documentation.

Here an outline execution plan and project time schedule for the later stages of the project are prepared.

Responsibilities and Documentation

In the study outlined in the previous section each member of the team has responsibilities to provide documentation, drawings, and specifications. Their roles are outlined in the following sections together with their detailed activities and the documents they must create.

Process Engineer

The process engineer of the contractor plays a key role in assisting the client to develop the optimum process to match his or her required production figures.

To develop the engineering required to achieve the design the process engineer will:

- Develop the process to the point where it can be communicated to the rest of the design team. This includes the formulation of each product, the specification of the active ingredients, and the excipients.
- Produce process flow diagrams
- Establish the utilities requirements from the conceptual process and equipment information and HVAC and dust control requirements.
- Calculate the utilities requirements for steam, compressed air, deionized water, purified water USP, towns water, vacuum, inert gases, etc.
- Develop utilities system designs.
- Interface with other team members to develop the materials handling, control, automation, and instrumentation concepts.
- Develop a validation philosophy to input into the design and prepare a draft validation documentation master plan.

Documents created

- Conceptual process design – design basis
- Process flowsheets
- Equipment list
- Preliminary piping and instrumentation diagrams for utilities
- Outline validation master plan.

Information required from client:

- Formulation for each product
- Specification of each ingredient
- Packaging specifications
- Process as listed in the Drug Master File and any subsequent changes or modification
- Any data available from the client's research and development and manufacturing group on systems already in use for products under consideration for manufacture in this plant.
- A forecast of capacity changes/product changes in the next five years including packaged stock.
- Any specialized equipment preferences, e.g. side-vented coating pans, fluidized bed coating columns, types of mixers, etc.

- The use of granulation end point control for wet granulation process within the client's group of companies.

Architectural

The *Guide to Good Pharmaceutical Manufacturing Practice, Food and Drug Administration* Section 21 requirements, relevant building regulations and The Fire Precautions Acts, will provide the design basis.

Scope of Work

- The conceptual layouts are developed to incorporate the following items:

    Clean/dirty spatial concepts
    Segregated clean/dirty personnel movement and changing accommodation philosophy
    HVAC and dust control philosophy
    Automated materials handling philosophy and logic flow
    Electrical/computer philosophy
    Maintenance requirements
    Plant room size and location
    Means of escape in case of fire
    Relationships with other facilities on the site.

- The design of the external appearance of the facility is developed but will not compromise the internal usage of the facility.
- Development of 1:100 scale drawings of the floor layouts, reflected ceiling layouts, sections, and elevations.
- Design of the external works adjacent to the facility.
- Development of preliminary room data sheets.
- Development of the fire compartmentation philosophy.
- Study of the finishes requirements for the facility.

Documents Created

- 1:100 scale floor layouts, sections, and elevations
- Preliminary room data sheets
- 1:1500 scale site plan.

Civil/Structural

It is assumed that basic soils and topographical data are provided by the client.

Engineering defines the following:

- Establish the frame type, slab type and clear span requirements of the facility
- Site preparation and earthworks requirements
- Preliminary foundation layout
- Frame design including main member sizes
- Slab design
- Preliminary drainage schemes for floor, surface, and chemical effluent
- Sketch of foundation layout and details
- Preliminary structural layout and main member sizes
- Preliminary drainage layouts.

Mechanical (HVAC/dust control, machinery, and piping). The engineering will define the following:

- Study to establish the HVAC/dust control and pneumatic handling requirements
- Schematics to demonstrate the HVAC/dust control, pneumatic handling and utilities systems
- Pipe and duct routing studies
- Definition of explosion hazards
- Control philosophies and interface with control and automation systems
- Location and sizing of plant rooms and services routing space requirements, particularly the ductwork within the overall facility layout
- Maintenance philosophy and space requirements
- Piping studies
- Schematics for HVAC and dust control
- Equipment list.

Electrical

The engineering will define the following:

- Analyse the KW ratings, variable speed drives, and all power supply requirements for the project, such as process equipment, HVAC, automated handling system, lighting, and small power.
- Nominally rate, scope, and physically size and locate the LV switchboards and MCCs.
- Review the essential loads and nominally rate the essential generator if required.
- Review the uninterruptible power supply requirements and nominally rate this equipment.
- Key one-line diagram showing the distribution from the Client's switchboards and MCCs, including the lighting and small power distribution boards.
- Location and equipment layouts for the switchrooms.
- Provide brief performance specifications for fire and alarm and detection systems, communication systems (e.g. Tannoy and telephone) and security systems.

Documents Created

- Key one-line diagram.

Control and Instrumentation

This section of the work establishes the application of computer techniques to provide 'state of the art' control and data processing to integrate the production, automated handling, environmental, and quality control requirements.

The design embraces the guidelines of cost, quality, safety, flexibility, security, loss prevention, and the CIM philosophy is developed, based on distributed computing techniques, and it covers the following task areas:

- Production scheduling
- Material control
- Recipe handling
- Process control and monitoring
- Shop floor automation
- Continuous quality control
- Data acquisition, monitoring, and processing
- Batch record documentation
- Management information
- Traceability and audit trail
- Archiving and retrieval
- Environmental control and monitoring
- Utility control and monitoring
- Building security.

The objectives of this study provide:

- A review of and a compilation of the user requirements
- A set of clear targets and acceptance criteria for the system
- An examination of alternative solutions and selection of a preferred approach
- A conceptual design in sufficient detail to confirm the technical approach and identify critical areas of system performance
- A statement of system requirement
- Recommendations on the resources and equipment required
- A review of financial and business justification and identify other benefits
- Development and implementation plans, including costs and timescales.

Documents Created

- Computer integrated manufacturing system conceptual design philosophy

- Development strategy
- Implementation plan.

Automated Materials Handling

The extent of materials handling work will depend on concepts developed during the process development stage. It is most likely that some form of automated materials handling system will be required. The work carried out will be a:

- Study to establish the scope of the automated materials handling requirements using the following basis:

  Automated movement of production 'IBCs' around the technical floor between dispensary, process units, QC, WIP, packing, and bin wash facilities. This interfaces with the control study to determine the level of monitoring and control that could interface with operator commands, a central work scheduling system, and quality control systems.

  Automated movement of packaging materials from the staging area to the packaging lines.

  Automated movement of packaged product from the packaging lines to the staging area.

  Layout criteria, physical size and load characteristics of the automated materials handling system to enable the layout to be finalized.

  Materials handling flow charts and layouts.

Documents Created

- Materials handling philosophy
- Outline equipment specifications
- Materials handling flowcharts and layouts.

3

Site Selection

Of course, the first thing to do was to make a grand survey of the country . . .

In deciding on a new site location for manufacturing a range of its products, a company has embarked on a considerable capital investment programme which will have a profound effect on its operations in the next 20 years. This decision is not easily reversed, and, therefore, it is essential to define the guidelines by which the new site will be selected. Usually the decision to look for a new location is based on a number of facts concerning the existing facilities. For instance, the existing site may be unsuitable for a variety of reasons such as:

- The production requirements have outgrown the capacity of the plant to expand to meet these needs.
- A completely new range of products will be manufactured.
- The site is outdated by the development of modern manufacturing systems and cGMP requirements.
- The site is located in the 'wrong' area, e.g. centre of a city, residential area, etc.
- The company has moved up market and wants to develop its public image.
- The local government planning authority will not grant further expansion for manufacturing on the existing site.
- The company wishes to embrace more computer controlled and monitoring systems to reduce its overheads and improve its costs effectiveness.

In this last *raison d'être* the company believes it can become more competitive by using systems such as computer integrated manufacturing (CIM), building management systems (BMS), and process monitoring and control systems (PMCS) to monitor and control the flow of materials, their use (improve yields) using fewer personnel, and reduce inventory levels.

Having arrived at the point where refurbishment of the existing facilities site is not possible then the company has to draw up a list of points that must be considered for the selection of a new site. Some of these are highlighted here. It is not a comprehensive list but more an indication of the type of decision to be made. The order of priority will differ depending on the particular requirements of the company. It is assumed that a green field site has been agreed on (as opposed to a new location using an existing building). So what steps need to be taken in selecting this new site? Probably the most fundamental is the development of a business plan which will define the objectives of the management team in the new location and the goals of the production staff (management by objectives, MBOs). It is essential to fully develop this plan at the start as many of the existing departments within the company will have differing priorities and interpretation of the requirements. Each group will provide a list of their needs. The resulting final project goals are the culmination of all these discussions and should be used by management to arrive at a final decision. Of course, events can always overtake the best decisions, so there has to be a built-in framework of flexibility.

Having arrived at the decision that a grass roots (green field) site is the only way forward, what factors must be considered in its selection? Most new projects of this nature commence from the viewpoint of a completely unrestrained budget (what would we like), and the next question is what will it cost? Only when this cost is addressed are the objectives trimmed to more realistic levels. In the selection of a new site there are 10–20 criteria that are of paramount importance. Some of these are considered here:

Economic Evaluation

Within the pharmaceutical industry there are a number of different types of company. There are the large multinational conglomerates (ethical) whose main products have certain common characteristics:

- They originate from research in their own laboratories.
- They originate from research conducted in universities.
- They have been licensed from another multinational. (In certain cases the introduction of a new product into a market can be achieved quicker by using an established sales team from an existing organization.)
- Their volume is low (high potency, may also be a sustained release product).
- They have good profit margins.

Few of these products are sold over the counter (OTCs) and thus rely on doctor's recommendation. Generally the OTCs have been separated into a subsidiary company owned by the multinational, and concentrate only on this type of product and its marketing. Returning to the case of the ethical company that wishes to relocate, it will be solely concerned with the economic facts of locating in a specific country and area.

What are the Tax Incentives?

Many governments eager to encourage industry offer sites with no local government taxes for the first five years of operation, give grants for buildings, and help with salaries over an initial period. Where a company is changing its location within a country's borders, then many authorities offer incentives to move into local unemployment black spots. Certain regional governments have developed scientific parks located near universities of technical excellence for prestige and high technology companies, and sites here are highly valued. A factor of increasing importance is the impact of environmental measures and regulations. It is vital to ensure that the local population is behind the siting of the new industry and that the company's effluent system fully conforms to local requirements. These costs are not small and are likely to grow as the 'green' lobby demands increase. Consider the case of Dow Chemicals in the Irish Republic who wished to build in a location near Cork on an area currently in use as a farm, but was forced to abandon the project owing to the environmentalist lobby.

Energy and construction costs can vary, depending on the location of the site. Some areas have access to low cost electricity (hydro-electric power), natural gas, or local solid fuels like oil and coal. Construction costs can be subjected to escalation outside levels of inflation owing to localized factors, e.g. building a new airport, railway, or the Channel Tunnel.

The technical evaluation of the facility will depend on a number of factors, i.e. labour intensive or capital intensive. A consideration of this will depend on where the new facility is being constructed. In western parts of the world, Europe, and the United States, there will be an emphasis on computer controlled systems, robotics, and low head counts. In areas of the world where labour is still cheap then traditional methods will be employed. However, if the company and the government controlling the local economy wish to export their products into the Western Healthcare system, then they will be required by that country's regulatory authority such as the MCA and FDA to ensure that their facilities and personnel meet certain minimum requirements for validation.

Chemical process plants require extensive effluent treatment systems, and the handling of waste for disposal and the method chosen is an important consideration. If, however, the centre is a distribution point, then transportation becomes a major factor.

Once the selection process has been narrowed to say two or three sites, then a topographical and geological survey of these should be undertaken to ensure that there are no hidden problems, e.g. old mine workings, hidden water courses, etc. Not only are these aspects important for the present facility site, but also for future expansion.

When a company has decided to move to a completely new location and has decided on the site from an economic viewpoint then the question of suitable staffing level must be addressed. Companies tend to locate in areas where there are a number of similar industries. This means that there will be a nucleus of trained personnel from whom it is usually possible to recruit. An added incentive is the possibility of a higher paid position and being part of a team developing a new facility from its conception.

One problem that presents itself in areas where the largest development grants are available is that the local workforce is mainly untrained in an industry such as pharmaceuticals. They generally come from a heavy industry background such as ship building or coal mining and regard light industry as being beneath their masculine dignity. Unions may also be a factor to consider. In addition many of the packaging operations require manual dexterity, and married women wishing to work part-time can be a valuable source of labour with this skill if there is a local town or housing complex.

Checklists provide a useful means of evaluating a site location and other factors associated with its selection. In the *Encyclopedia of Chemical Technology* an extensive checklist is provided (Volume 18, p. 45).

4

Process Flow

And she took a ribbon out of her pocket, marked in inches, and began measuring the ground, and sticking little pegs in here and there.
'At the end of two yards ...'

The basic criterion from which any plant is designed is an understanding of the process, the equipment, material flow, and people flow through the facility. There are three documents that need to be produced at the early stage of the development of the facility. These are process flow diagrams (PFD) (Figure 4.1) and two additional documents which include engineering design criteria and information relating to plant operation and safety.

A process flow diagram is developed from direct knowledge of the project and partly from information supplied by the sales and marketing departments of the company requiring the new facility. The intent is to define the process philosophy and process design criteria to best meet the immediate and future requirements of the product. Their main purpose, as of any drawing, is to communicate information in a simple and explicit way, using unscaled drawings which describe the process. Sufficient detail must be presented on the PFD to give any experienced process engineer an adequate understanding of the process concepts, operating conditions, and equipment sizes, to permit a critical review of the process design with minimum reference to other documents, i.e. process description. Process diagrams and piping and instrumentation diagrams (P and IDs) (Figure 4.2) each have their own functions and should show only information that is relevant to their particular needs. Extraneous information such as piping, structural and mechanical notes should not be included unless essential to the proper performance of the process design. Instrument control and manual control valves which are necessary for the operation of the process are shown together with instrumentation essential to process control. Systems for providing services are not shown; however, the type of service, flow rates, temperatures, and pressures is noted at consumption rates corresponding to the material balance. Separate flow diagrams are preferred for each utility system, such as clean steam, purified water, compressed air, heating, ventilating and air conditioning (HVAC), and specialized gas systems. These are an integral part of the process. The diagrams show all items of equipment connected to the systems in continuous use and show consumption levels, capacity levels, and specifications for the requirements of the air and air flow volume In addition the routes of disposal for the effluents are shown.

Each PFD has a separate set of tables which show the quantitative and qualit-

BASIS
1 g direct compression coated tablet
1000 mg (600 mg active 400 mg excipient)
Aqueous film coating suspension/solution
Solvent film coating suspension/solution
650 million = 650 tonnes = 2.955 million per day   3 tonnes per day
60% active = 1.8 tonnes per day
40% excipient = 1.2 tonnes per day

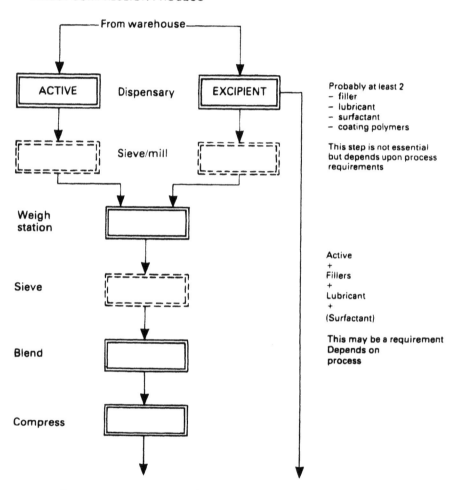

DIRECT COMPRESSION PROCESS

Figure 4.1 Product A: Requirements 650 million per annum. 1 g (1000 mg) direct compression core, film coated tablet can use either an aqueous film coating suspension/solution or a solvent film coating suspension/solution. 600 mg active 400 mg excipient

ative data for any process stream. The extent of data is the minimum required for sizing the lines and the development of the P and IDs from the PFD.

In addition to the heat and material balances the working documents need to be developed from the PFDs to record the engineering operating and safety criteria of the plant design. These documents record the engineering design criteria and are maintained until the contract reaches the approved for design (AFD) stage. The second working document is the repository of all matters relating to plant operation

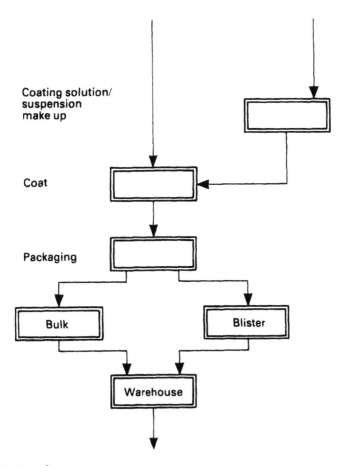

Coating solution/
suspension
make up

Coat

Packaging

Bulk

Blister

Warehouse

Figure 4.1 (continued)

and safety and hence is maintained until an operating manual is issued. Both of these documents are prepared by adding the necessary information to those parts of the process flow diagrams. The purpose of the first working document (WD) is to define the maximum operating limits throughout the plant and ensure that these limits are applied consistently item by item. The WD initiates work on the process data sheets for the equipment and the P and ID (a typical P and ID (engineering line diagram–ELD) is shown in Figure 4.2).

The second working document describes the basic operating requirements which are to be incorporated into the plant design. The purpose of this document is to ensure that the interactive nature of the total plant is thought through carefully and established before downstream design work starts. This is particularly important on multi-unit plants. This document, therefore, describes the operating philosophy and indicates all the additional process lines and any associated control loops required, their functions and sizing basis.

For convenience, various flow sheets and process flow diagrams for the manufacture of various types of tablet are illustrated in the next few pages. These flow patterns give an indication of the materials and size of the equipment required at each unit operation. Also shown are material flow diagrams for a number of different building configurations. In these diagrams the basis for production purposes is a single 7 hour shift working; 5 days per week; 220 days per annum.

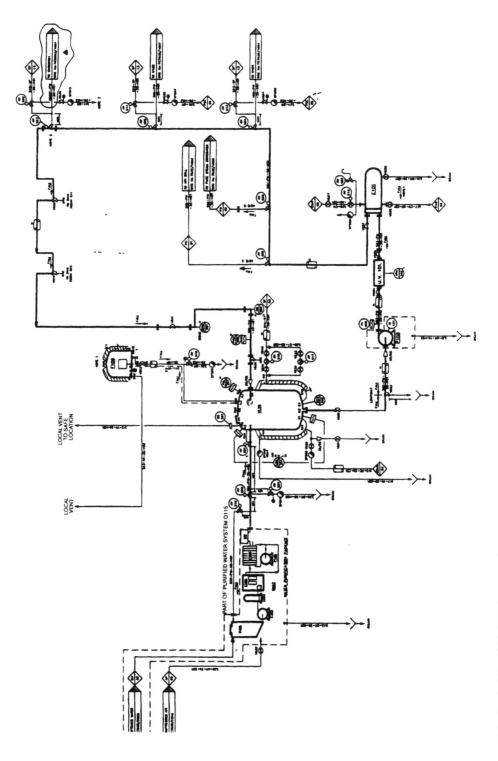

Figure 4.2 Typical P and ID (ELD) for part of a purified water system

Batch size will depend upon the process.

For convenience, 500 kg will be used as a sub-batch and 6 sub-batches will be consolidated in one day's production, to give one batch of 3000 kg (3 tonnes) for analytical control purposes.

Storage capacity will depend upon the inventory control philosophy. Here, one week (5 days) of raw materials is used as the design basis.

Therefore, sufficient capacity to store 9000 kg of active and 6000 kg of excipient material should be stored in drums or bulk containers. Identification of individual batches is essential. Bulk containers (IBCs) should be considered for the storage of the active material and the main excipients, and the design of the IBC would be based on the size of the batch and the material characteristics. These IBCs could be used directly in the process or the material transferred into a bulk storage silo adjacent to the dispensary.

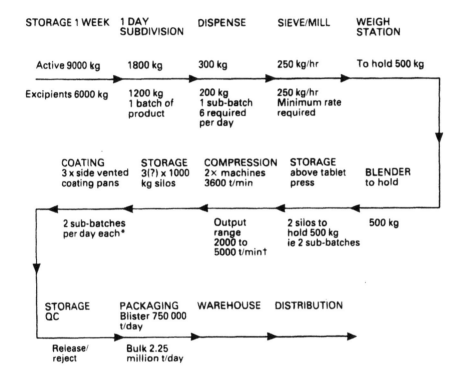

* Will depend upon the process. Assume 2.5 hours to coat, 0.45 hours to load, discharge and polish, giving 2 sub-batches of 500 kg per day.

† 3 million tablets required per day = 2 x tablet presses capable of 3600 tablet per minute output.

Figure 4.3 Product A: Requirements 650 million per annum. Direct compressed core, film coated tablet weighing 1.0 g

Legend

DCT: Direct compression tablet
CT: Compressed tablet
FCT: Film coated tablet
→ : Process flow
IBC: Intermediate bulk container
Light grey, dark grey, black: These classifications represent various levels of cleanliness ranging from white for a sterile suite to black for unclassified.

This information can be illustrated to provide additional data on batch sizes and equipment requirements (see Figure 4.3).

Batch size will depend upon the process.

For convenience, 500 kg will be used as a sub-batch, and 6 sub-batches will be consolidated in one day's production, to give one batch of 3000 kg (3 tonnes) for analytical control purposes.

Storage capacity will depend upon the inventory control philosophy. Here, one week (5 days) of raw materials is used as the design basis.

Therefore, sufficient capacity to store 9000 kg of active and 6000 kg of excipient material in drums or bulk containers is required. Identification of individual batches

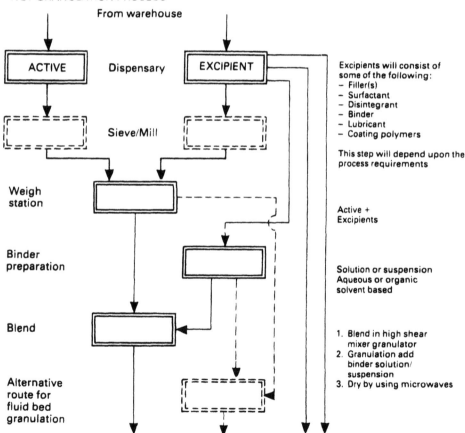

Figure 4.4 Product B: Requirements 300 million per annum. 200 mg FCT prepared using wet granulation, can use either an aqueous film coating or a solvent film coating. 2 mg active, 198 mg excipients

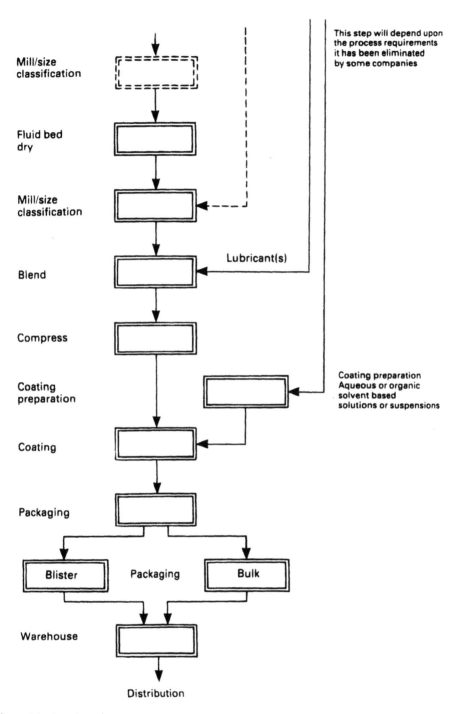

Figure 4.4 (continued)

is essential. Bulk containers (IBCs) should be considered for the storage of the active material and the main excipients, and design of the IBC would be based on the size of the batch and the material characteristics. These IBCs could be used directly in the process or the material transferred into a bulk storage silo adjacent to the dispensary.

Alternatively, data required for a compressed film-coated tablet using wet granulation methods is shown in Figure 4.4. Figure 4.4 can be improved to give a spatial

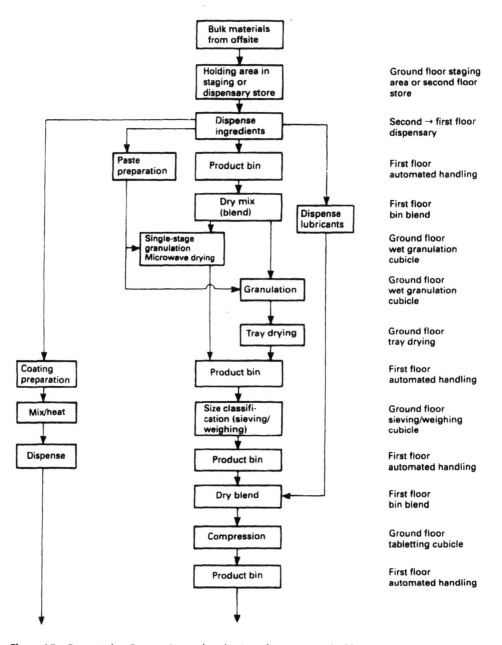

Figure 4.5   Process plan. Preparation and packaging of sugar-coated tablets using wet granulated methods

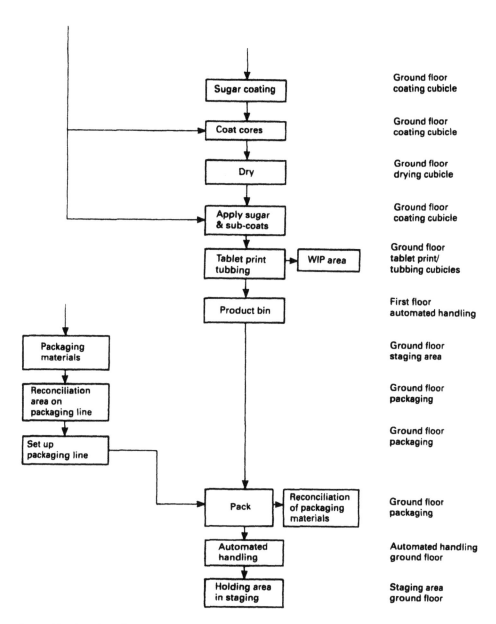

Figure 4.5 (continued)

appreciation of how each unit operating links together and the areas that are required. This is shown in Figure 4.5.

In the following flow diagrams the following legend is used: black, dark grey, and light grey. White would be used to signify a sterile area but in the examples shown, it is not applicable. Black signifies the basic requirement which might be used for a GMP warehouse. Dark grey would be an increase in standard which might apply to a packaging area and light grey may be used in a solid dosage manufacture area to

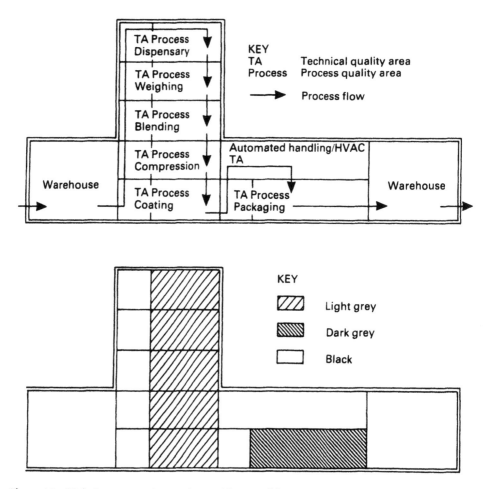

Figure 4.6 High rise tower using gravity to aid material flow

indicate an improvement in the finishes and quality of air that the HVAC is designed to provide.

In Figures 4.6 to 4.9 two diagrams are shown: the diagram is a schematic section through the building indicating process flow and the bottom diagram is a schematic footprint indicating the types of finishes that have to be considered for the various environmental requirements. These figures represent differing options for the flow of materials through various building arrangements.

The transfer of the product from coating to packaging can use a pneumatic transfer system, a conveyor, or an AGV/IBC bulk container for feeding the packaging line.

Option A (Figure 4.6)
This building has four upper floors and a ground floor designed to maximize the use of gravity in feeding powders and product starting in the dispensary and finishing in the packaging area.

Option B (Figure 4.7)
This is designed to maximize the use of pneumatic/vacuum transfer systems and is intended to be a totally enclosed dedicated system. There would be an option within

the design to change the transfer system from each unit of equipment for each

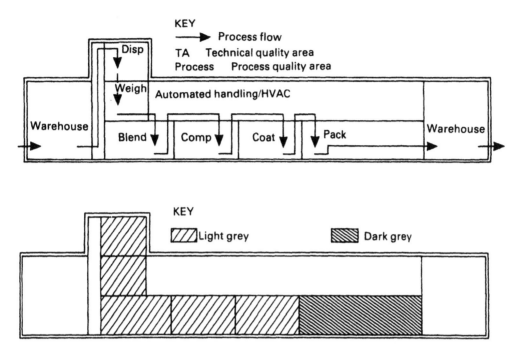

Figure 4.7   Three floor system using automated guided vehicles to aid material flow

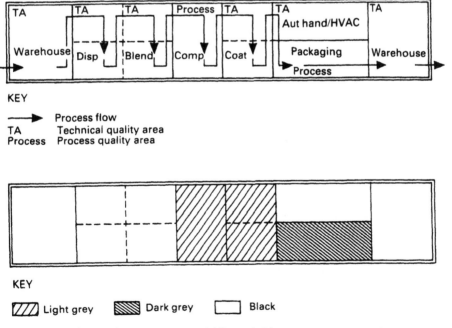

Figure 4.8   Single floor with mezzanine material flow aided by vacuum or pneumatic systems

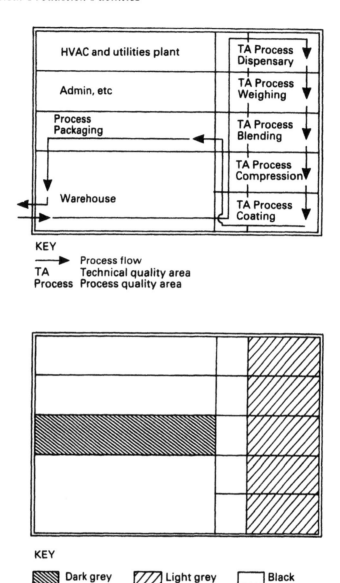

KEY
→ Process flow
TA   Technical quality area
Process Process quality area

KEY
▨ Dark grey   ▧ Light grey   ☐ Black

Figure 4.9 High rise tower system incorporating warehouse, utilities, and administration

product to operate the plant on the campaign basis.

Option C (Figure 4.8)
This is the 'Lhoest' concept; the pneumatic transfer systems have been replaced by an AGV/IBC mechanical handling system.

Option D (Figure 4.9)
This is similar to option A with the exception that the process route has been so designed to minimize the product and people flow and use one warehouse for incoming raw materials and outgoing finished product.

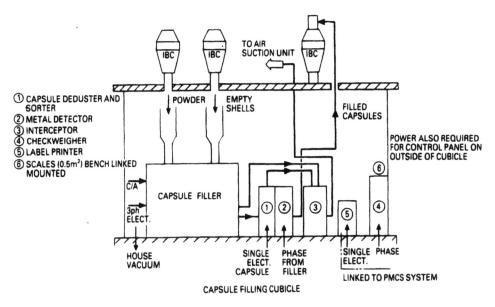

Figure 4.10 Capsule filling

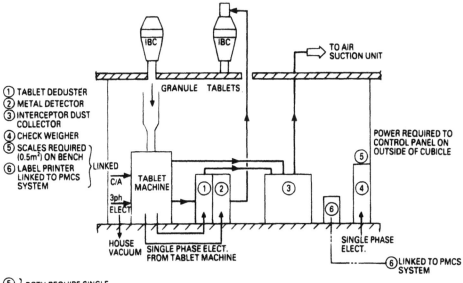

Figure 4.11 Tablet compressing

Figures 4.10 to 4.13 represent some examples of PFDs for tablet compressing, capsule filling and coating cubicles using an IBC as the vehicle for material transfer.

The flow diagrams (Figures 4.1 and 4.4) shown for product A and product B illustrate a direct compression process and a wet granulation process, and are intended only to highlight some possible options. Figure 4.3 shows the mass flow of material and product through a facility. These have been developed by using company data and making a number of assumptions. These are:

41

- The tablets can be compressed, coated, and packaged in a normal manufacturing environment controlled at 20°C ± 2°C and 50% RH ± 10% RH.

- The tablets are round and bi-convex (i.e. no shapes). They may be intagliated with a product number, symbol, logo, or break bar.

- Sugar coating is not required.

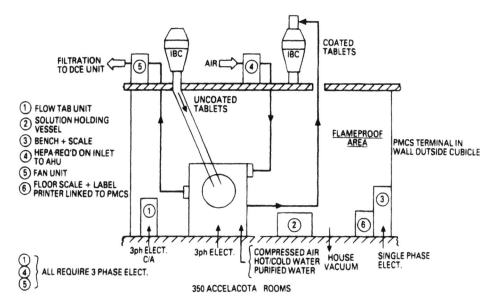

Figure 4.12 Film coating

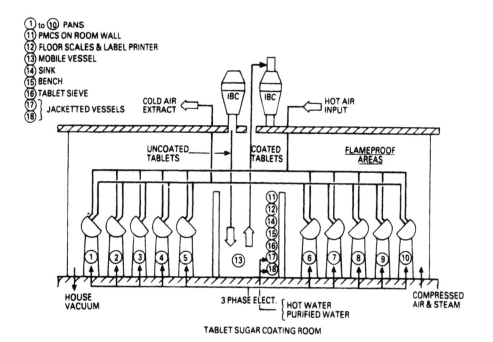

Figure 4.13 Sugar coating

42

Equipment data sheet	Project No:
	Rev:

Equipment:	Equipt ref:
Manufacturer:	Model:
Tel No: Contact:	Location:

ELECTRICAL SERVICES

Voltage	V	Service Termination:	Panel/Isolator/Socket
Max. Power	KW	Computer Link:	Yes/No
FLC	AMP	RCD Protection:	Yes/No
Special Requirements:			

MECHANICAL SERVICES

SERVICE	PRESSURE	FLOW	DAILY CONSUMPTION	CONNECTION SIZE
Domestic Hot Water	(Bar g)	(kg/hr)	(L)	(mm)
Tank Cold Water	(Bar g)	(kg/hr)	(L)	(mm)
Mains Cold Water	(Bar g)	(kg/hr)	(L)	(mm)
Softened Water	(Bar g)	(kg/hr)	(L)	(mm)
Purified Water	(Bar g)	(kg/hr)	(L)	(mm)
WFI	(Bar g)	(kg/hr)	(L)	(mm)
Drain	N/A	(kg/hr)	(L)	(mm)
Service CA	(Bar g)	(L/S)	N/A	(mm)
Instrument CA	(Bar g)	(L/S)	N/A	(mm)
Nitrogen	(Bar g)	(L/S)	(L)	(mm)
Oxygen	(Bar g)	(L/S)	(L)	(mm)
Argon	(Bar g)	(L/S)	(L)	(mm)
Propane	(Bar g)	(L/S)	(L)	(mm)
Natural Gas	(Bar g)	(L/S)	(L)	(mm)
Vacuum	(Bar Abs)	(L/S)	N/A	(mm)
Dust Extract	(Pa)	(m³/s)	N/A	(mm)
Steam	(Bar g)	(kg/hr)	N/A	(mm)
Clean Steam	(Bar g)	(kg/hr)	N/A	(mm)
Condensate	N/A	(kg/hr)	N/A	(mm)
Vent	N/A	N/A	N/A	(mm)
Cooling Water Temp °C	(Bar g)	(kg/hr)		(mm)
Special Requirements:				

Figure 4.14 Equipment data sheet

Additional documents that are very useful at the early stage of any project are room data sheets (Figure 4.15, pp. 44–45), equipment lists and equipment data sheets (Figure 4.14, p. 43).

Room data sheet	Project No:
	Rev:

Project:	Date:
Client:	Page:
ROOM NAME	ROOM No
Function	Occupancy
Area (SM)	Plan (M)	Height (M)

ENVIRONMENT

Standard	Class level
Temperature (°C)	Humidity (%)
Hazards	Class level

AIR SUPPLY

Changes (/hr)	Forced/natural
Pressure (Pa)	Air Pattern
Filtration	
Recirculation (%)	Heat recovery (%)

FINISHES

Walls	
Doors	
Windows	
Ceiling	
Floor	Coving

DRAINAGE

Floor outlets
Sinks/equipment
Spills/washdown

ELECTRICAL

Power outlets	
Lighting level (lux)	Fitting type

COMMUNICATIONS/FIRE PROTECTION

Phone	Data link	Clock
Fire detection		Alarm
Sprinklers		

Figure 4.15   Room data sheet

PROCESS UTILITIES

WATER		
Mains cold water	Tank cold water	Hot water
Deionised	Distilled	WFI
Steam	Clean system	Condensate
Chilled water	Glycol	
COMPRESSED AIR		
Plant air	Process air	Exhaust
GASES		
CO2	Helium	Hydrogen
Natural gas	Nitrogen	Oxygen
OTHER PIPED SERVICES		
Vacuum	Vacuum cleaning	

PLANNING DATA

LOCATION
Rooms adjoining
Rooms nearby
Service zones
ACCESS
Normal
Emergency
Maintenance
Clothing

EQUIPMENT

Equipment reference	Data sheet no.

AUTHORIZATION Name	Date	Revisions				

Figure 4.15 (continued)

5

Pharmaceutical Process
Utility Systems

As she said these words her foot slipped, and in another moment, splash! she was up to her chin in salt-water.

Process utility systems are designed to satisfy the requirements of the facility, and are based on the operational philosophy, i.e. single shift, double shift, or 24 hour working. The capacity requirements of the various utility processes and process equipment are estimated by using existing in-house data and data from equipment vendors, and allowing for expansion in the future. A contingency factor must be applied to the design margin to accommodate occasional peak demand or for uncertainties when designing equipment.

All facilities will require the following services:

- Heating, ventilating, and air conditioning (HVAC)
- Hot and cold water
- Chiller package
- Effluent treatment
- Steam
- Electrical services
- Compressed air
- Vacuum systems
- Dust collection.

The design criteria will depend to a large extent on the quality of service required.

In addition, special services will be required, depending on the types of product being produced, for example:

- Water for Injection USP
- Purified Water USP
- Clean steam
- Nitrogen or other inert gases
- Breathing air/instrument air
- CIP system.

Sound engineering design is only part of the equation in creating a system to produce the quality of product required. It also requires proper management in operating the equipment. This is where validation provides the necessary design review criteria and the experimental design for operating and maintaining the quality of the product. The supply of pure water through a piping network can best be assumed by following a number of validation procedures during installation. Some of the factors that need to be evaluated in the design will be considered here, using purified water, water for injection, HVAC and dust collection and compressed air as examples.

The pharmaceutical manufacturer is faced with the continual need to review and upgrade his water systems. He must ensure an adequate supply of water that will satisfy all criteria of quality, quantity, and reliability. There is a good deal of diversity of opinion regarding desirable water quality and the means of satisfying certain requirements. This is caused by the constantly changing technological advances being made, and the economic overtones which dictate a need for more 'cost-effective' systems.

The pharmaceutical industry comes under the auspices and regulations of the FDA and MCA, and they are reviewing with increasingly greater attention the treatment and handling of water used for washing, rinsing, and product formulation in all segments of the industry.

The area of concern here is the need for quality water in the preparation and manufacture of these products:

- Parenteral drugs (LVPs and SVPs)
- Biological drugs
- Over the counter drugs (OTC)
- Diagnostic drugs
- Radiopharmaceuticals
- Veterinary
- Medical devices.

Generally the two grades commonly used are Purified Water USP and Water for Injection (WFI).

Water is one of the most difficult products to maintain to the standard required especially as the quality increases. Not only do the chemical impurities (Table 5.1) require treatment, but it is also necessary to consider microbiological contamination. This is because systems contain components that support and encourage bacterial growth. Some examples of systems design are illustrated here for the more common process utility services, together with some suggestions and ideas on improving the engineering design. It is not an exhaustive list but more of a road map to lead both user and designer to achieve the required goal.

Purified Water USP (PW) must contain no added substance, and therefore microbiological control of this water is difficult unless it is handled in a manner similar to that of Water for Injection (WFI). The use of this class of water should therefore be restricted to those operations that mandate the use of PW. Many operations can be performed by using potable water, and do not need the high chemical and microbiological quality of the more difficult to control PW systems.

PW, since it is difficult to control from a microbial standpoint, should be used for critical bulk batch applications – where there is no reasonable alternative – and non-sterile, non-parenteral product formulations. For existing facilities, where

Table 5.1 Quantitative interpretation of USP

	Purified water standards
Test	Required Standard
Colour	Colourless
Odour	Odourless
Appearance	Clear
Specific resistance	100 000 ohms-cm minimum
pH	5.0–7.0
Chloride	0.5 ppm max
Sulphate	0.5 ppm max
Ammonia	0.3 ppm max
Calcium	0.5 ppm max
Carbon dioxide	4.0 ppm max
Heavy metals	0.5 ppm
Iron	0.1 ppm
Copper	0.01 ppm
Chromium	0.01 ppm
Cobalt	0.1 ppm
Manganese	0.1 ppm
Nickel	0.1 ppm
Oxidizable substances	Meets test
Total solids	10 ppm max
Microbiology	Not more than one colony forming unit per 100 ml (as a mean of all samples examined per month). Free from lactose fermenting organisms, yeasts, and moulds

deionization equipment (DI) is installed, the PW can be produced by DI followed by distillation, reverse osmosis (RO), or ultrafiltration (U/F). The still, RO, or U/F unit, storage tank and piping distribution system should be periodically treated by an appropriate means to control microbial growth. For new facilities, methods of treatment may be DI followed by preferably distillation – RO or U/F may also be used, following the ion-exchange equipment, depending on individual considerations of flow, volume, feedwater quality, and economics.

In all solid dosage, oral liquid and creams manufacturing facilities, Purified Water USP (Table 5.1) is used for in-process manufacturing and as a final rinse when cleaning equipment. The following is a description of one such system which generally consists of the following units:

- Feed water system
- Pretreatment system
- Deionizer system and recirculation loop
- Filtration system
- Storage and distribution system.

The purpose of this process is to convert town water into water that complies with the standard defined in the *US Pharmacopeia* as Purified Water USP. Purified water is prepared by deionization from feed water obtained from the municipal water supply.

The system consists of pretreatment followed by a treatment system composed of a twin bed deionizer, mixed bed deionizer, a polisher, a UV unit, and 0.2 μm filter prior to the entry of the water into the storage tank. The treated water is constantly recirculated from the tank through the ambient temperature distribution loop to the point of use outlets and back to the storage tank. A typical system is illustrated in Figure 5.1.

Feedwater system Feedwater is fed to the system from a polypropylene break tank, where capacity depends on the design requirements of the system. The flow is initiated by a service pump activated by high and low level probes in the treated water storage tank water. When the water level is low, a signal from the controller activates the service pump and initiates flow through the entire system. A low-level probe in the break tank is linked to the service pump through a controller to sound an alarm and switch off the pump if a low-level water condition exists.

Pretreatment system From the break tank, the water flows through a pretreatment system consisting of a multi-media filter, an organic trap, and a carbon filter. The multi-media filter is composed of granular anthracite, monograde filter sand, and gravel. The multi-media filter removes both total solids and colloidal material from the incoming water. It is backwashed by using the infeed water every two days. The backwash cycle is timer controlled. The organic trap contains a resin such as Rohm + Haas IRA958 which removes organic material that would otherwise foul one anion bed of the twin bed deionizer. The organic trap is regenerated with a salt solution. Regeneration is initiated by a timer on the unit or from a flow meter which activates the regeneration cycle after a definite quantity of water has passed or by a combination of these controls which are dependent on time, flow, and quality of water. The salt solution is fed from a tank located adjacent to the unit.

The carbon filter removes residual chlorine along with some organics. A typical activated charcoal bed uses Chemviron Filtrasorb 200 with a gravel underfill. The backwash of the filter bed is usually timer controlled, but can be controlled by a similar means to the organic trap. The backwash water source is the infeed water to the system.

The pretreatment units are connected by piping which uses typically an ABS (acrylonitrite, butadiene styrene compound) resin. These sections of piping are joined by bonding or by a threaded pipe with a collar connection. There are gate type sampling ports installed in the piping before and after each of the filters.

Deionization system and recirculation loop After pretreatment, the water passes through a twin bed deionizer containing typically a cation resin, IR120, and an anion resin IRA458. The unit has a fully automatic regeneration control circuit, and its output is monitored by an insertion type conductivity cell. If the conductivity meter exceeds a preset maximum of say 5 microsiemens for the output water, then the deionizer automatically initiates a pre-rinse cycle and, if necessary, a full regeneration cycle.

The cation resin is regenerated with a dilution of 30% HCl solution obtained from a tank adjacent to the cation filter and the anion resin is regenerated with a dilution of 30% sodium hydroxide solution obtained from a tank, both being controlled by pneumatically operated valves. The regeneration sequence begins with a cation bed backwash, followed by dilute acid injection and a rinse and then followed by the anion bed backwash, dilute caustic injection, and rinse. When complete flow

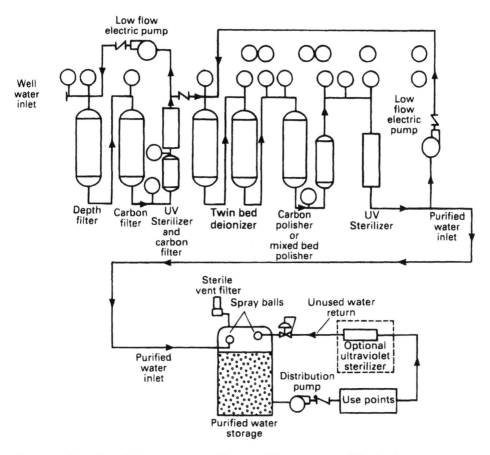

Figure 5.1 Typical purified water system with recirculation, storage and distribution system

to the system is restored the system is designed to regenerate after a finite quantity of water has been treated. If the design flow rate is between 500–13 00L per hour, then the twin bed deionizers should regenerate approximately every 19–50 hours assuming consistent pretreatment water quality.

The water for the twin bed deionizer regeneration rinse is fed from the break tank by the service and regeneration pumps, which are activated when the system starts the regeneration pre-rinse cycle. Regeneration water is filtered through a 5.0 μm activated carbon filter. The water which flows through the service pump is filtered through the pretreatment system. The service and regeneration lines join to form a common inlet to the deionizer. The water from the twin bed deionizer flows through a mixed bed deionizer or polisher. The resistivity of the water entering the polisher should be in the region of 0.2 megohms. The water is polished to 1–10 megohms by this unit. The water purity is monitored by an analogue resistivity indicator/ ~ntroller, whose output is recorded as part of the energy management system data ; provide recorded data on the purified water quality. This data is also sent to the water central control panel. When resistivity output level reaches the alarm condition, i.e. 1 megohm, the inlet solenoid valve to the storage tank is closed and a dump drain valve will open. The water will be diverted to drain as long as the resistivity is out of the specification range.

Filtration loop On leaving the polishing unit, the water flows through a 0.45 μm filter, a UV unit and a 0.2 μm filter. The 0.45 μm filter is contained in a single element 20 inch 316 stainless steel filter holder. This cartridge is a polypropylene filter with a nominal pore size of 0.45 μm. The intensity monitor uses UV light at 254 μm for exposure of the circulating water, and the 0.2 μm filter holder is a stainless steel electropolished 30 inch single element housing, containing a cellulose acetate cartridge filter with an absolute pore size of 0.2 μm.

After filtration, the treated purified water is stored prior to being fed into the distribution loop and the points of use. At the storage tank the flow is separated into two streams, using the tank inlet valve to divert the flow of water. Two thirds of the treated water enters the storage tank and one third is recirculated through the twin bed deionizer. The deionizer loop contains a pump which generates heat in the loop. To monitor the temperature of the water a sensor is located downstream of the pump. When the operating temperature of the water in the loop is exceeded, the dump drain valve located before the inlet of the treated water storage tank opens allowing treated water to enter the recirculation loop. Once the water temperature has been lowered, the drain valve closes.

Located downstream of the recirculating pump is a pressure sensor, which operates when the system pressure exceeds a preset limit. The dump drain valve then opens to release the excess pressure. When the pressure is reduced, the dump drain valve closes.

The piping used in the deionization system and the regeneration loop can be composed of Durapipe ABS resin. The connections are either bonded with adhesive or joined by a threaded collar connection method similar to that used for the pretreatment components.

Storage and distribution system The treated water storage tank is a suitably sized tank constructed out of 316 stainless steel with an electropolished interior designed to hold at least one shift's usage, but will depend on the operating philosophy of the plant. It is fitted with a manhole opening, pressure relief valve, and vent filter located on top of the tank. The vent filter is a sterile cartridge type 0.2 μm polypropylene filter. Water at ambient temperature from the tank is circulated by a sanitary type distribution pump through a heat exchanger, UV unit, and 0.2 μm filter around the distribution piping to the points of use and returns to the storage tank. The heat exchanger is for use in the sanitization of the distribution system when required. The UV unit is an intensity monitor which uses UV at 254 μm as the exposure frequency. The 0.2 μm filter consists of multi-element filter housing, constructed of electropolished 316 stainless steel containing 8–30 inch cartridge filter elements with a 0.8 μm nominal filter preceding an absolute 0.2 μm cellulose acetate layer.

The distribution loop is constructed from 316 stainless steel electropolished piping. There are sanitary diaphragm sampling ports installed at the outlet of the storage tank and at the distribution loop return to the storage tank. The piping is welded by Argon arc TIG welding.

High- and low-level sensors are fitted to the treated water storage tank, and these sensors control the inlet solenoid valve of the tank. If water is required, the valve opens to allow water to flow into the tank. Alternatively, if the level in the tank is high the tank inlet valve closes. The level sensors in the tank also turn off the distribution pump to avoid damage to the pump if the water level is too low.

The recirculation flow rate is designed to produce turbulent flow in the piping to reduce formation of a bio-layer on the piping interior. In the event that the system requires sanitization, the heat exchanger will be used to heat the circulating water to 85–90°C. After sanitization, the system temperature will again be reduced to the ambient operating temperature range.

The design of this type of system lends itself to the proliferation of bacteria. The multi-media filter probably contributes least to this increase, but the carbon filter removes free chlorine, and residual chlorine stops bacterial growth. Any absorbed organic contaminant in the carbon filter provides nutrients for bacteria growth. In the deionizers, the absence of chlorine and the presence of a source and nutrients support bacteria growth. However, the use of hydrochloric acid and caustic solution during the regeneration sequence upsets the ecology balance for the bacteria, thus providing a sterilization process as a byproduct and reducing bacteria growth. When storage tanks are used it is essential that proper management of the distribution loops are provided to supplement the design of the system, i.e. correct welding techniques, passivation of the system, absence of dead legs.

Alternative systems using reverse osmosis and electrodeionization are shown in Figures 5.2 and 5.3.

Water for Injection (WFI)

Water for Injection is the most difficult quality to achieve and must satisfy the specifications for Water for Injection as defined by USP XXI. There are stringent chemical as well as microbiological and pyrogenic requirements. The water must be chloride free, and contain no added substance, which precludes the use of residual chlorine as a microbial control agent. A typical process flow sheet is shown in Figure 5.4.

There are two acceptable methods of production in USP XXI; distillation and RO. To produce USP Water for Injection a double pass RO system must be used to ensure any degree of consistent water quality, especially in the biological and pyrogenic species areas. All membrane devices have seals and 'O' rings, and during periods of shutdown these sealing materials will relax and minute quantities of bacteria, pyrogens, etc., may leak through. In addition, if tiny pinholes developed in any membrane configuration a double pass system would ensure added protection.

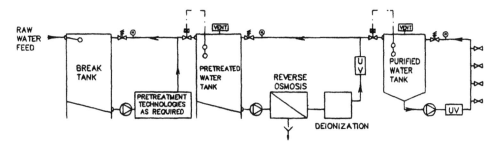

Figure 5.2   Basic purified water system design

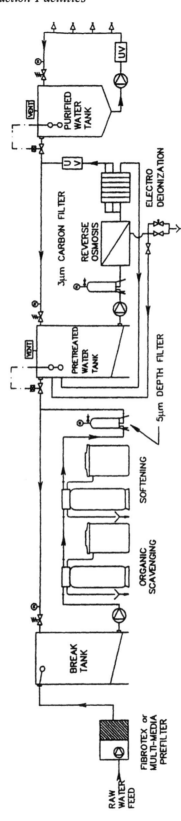

Figure 5.3 Pretreatment/RO/CDI purified water system design

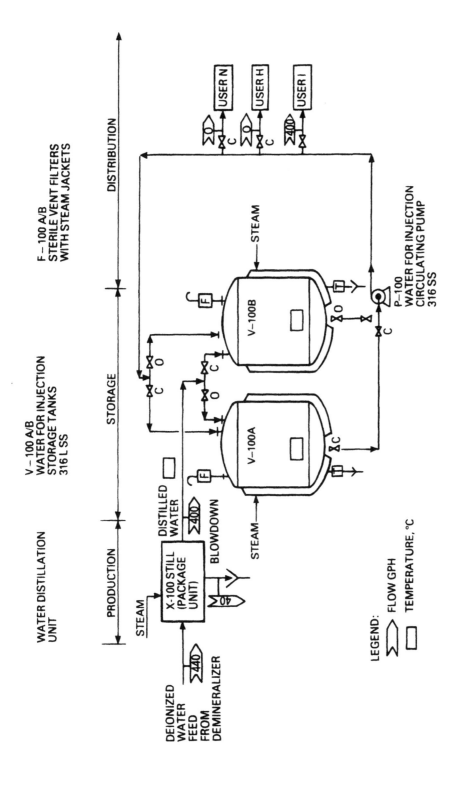

Figure 5.4 Process flow sheet for water for injection system

55

Careful attention must be paid to the selection of a type of membrane material and membrane configuration. This is due to variations in feedwater conditions, and the complex ·pretreatment system that may be required to protect the membrane surface and an adequate quality and quantity with system reliability.

It is for these reasons, and the complex pretreatment required for RO, because of the membrane's sensitive nature, that distillation is the more accepted and sound approach for WFI water.

Existing systems that distil at elevated temperatures but do not continuously recirculate and maintain 80°C will be acceptable if the water is discarded daily. At the use point, hot water flushing procedure must be validated and monitored to ensure that the water is consistently within microbial specification.

New systems should be designed to produce hot distilled water continually recirculated at elevated temperatures. WFI should be used hot whenever possible. At those use points where the water cannot be used hot, it should be cooled only while actually using the water, and the system design must ensure that water that has been cooled is not returned to the storage tank.

Figure 5.5 illustrates a typical distribution scheme for cool distilled water. The individual use point shown is provided with a shut-off valve close to the main header. The design should incorporate a provision for adequate flushing of the branch line by automatic means or by following carefully written manual procedures.

Multiple use points (see Figure 5.5) within a close physical area can be fed by a sub-loop that is cooled. Suitable automatic controls must be selected to ensure that the outlet valve of the loop will be closed at all times when the water in the loop is below the system set temperature.

In design today, all pipes are sloped to provide self-draining, dead legs are avoided, and threaded fittings have been replaced by fusion welds which ensure an almost seamless joint, thus reducing potential contamination sites. The importance of these engineering specifications in maintaining control of the system quality cannot be over-emphasized.

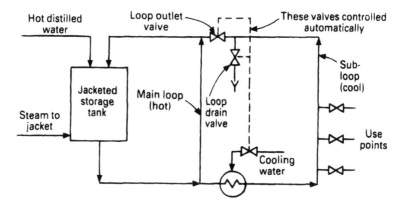

Figure 5.5 Multiple cool water use points. When cold water is not needed, water will circulate through the main loop and the sub-loop at 80°C. During use, the cooling water valve is open and the loop outlet valve is closed. After use, the loop outlet valve is kept closed and the loop drain valve opened to flush all the cooled water to drain. This will prevent any water that has been cooled from being returned to the storage tank. Heat exchanger design will comply with the proposed LVP GMPs

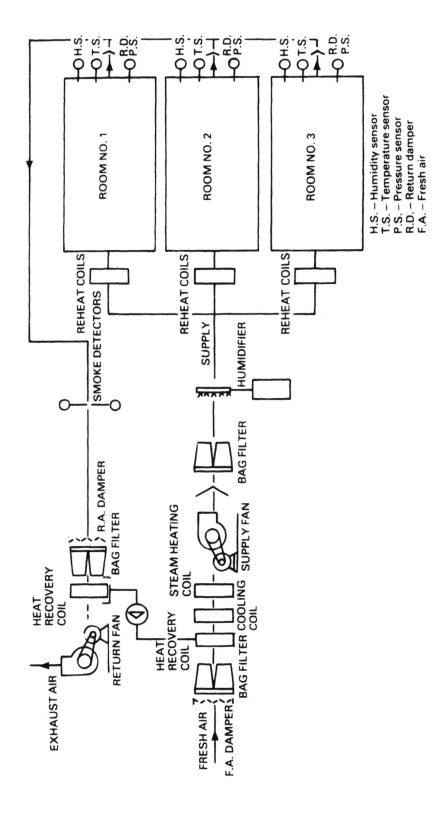

Figure 5.6 Typical air conditioning layout

Heating, Ventilating and Air Conditioning (HVAC)

A typical layout of an air conditioned plant is shown in Figure 5.6. Various types of systems can be installed to provide:

- Constant volume, variable temperature
- Constant temperature, variable volume
- Split systems
- Variable volume and temperature
- Cooling and dehumidifying, using chilled water
- Heating systems – steam, water, electrical
- Filtered air to 99.99%
- Humidifying, using direct heat injection, ultrasonic humidifiers, and water bath
- Pressurization, using system balancing or automatic dampers
- Heat recovery, using thermal wheels, air to air recuperators, heat pipes, or water to air systems.

The complexity of the systems will depend on the requirements within the plant. The question of filtration has to be considered in view of the standards of particulates in the air. Is the area required to be Class 100, 10 000, or 100 000? Is HEPA filtration required? What problems of cross-contamination exist? HVAC is a very expensive part of the process utilities, and the air volumes need to be kept to a minimum to make efficient use of the energy. Air flow patterns must be considered in view of the heat gain provided by process equipment. Humidity and temperature control of the air must be addressed; some effervescent products require very low humidities (less than 20% RH at 20°C) and some products are very deliquescent.

Figure 5.7 shows how heat gain from a unit of equipment affects temperature within a room.

Dust Collection

The removal of airborne particles in any pharmaceutical facility is essential for a variety of reasons:

- They present health hazards
- They contaminate the local environment and/or other products
- They can be highly toxic
- They are potential sources of explosions.

In many modern facilities the processing of materials, i.e. milling, and granulation are contained in an area which provides a barrier between the operator and the product being manufactured. Many new compounds are highly potent, and it is necessary to ensure that the product does not contaminate the operator and vice versa. The *Orange Guide* states:

> The handling of dry materials and products creates problems of dust control and cross contamination, and special attention is needed in the design, maintenance, and use of

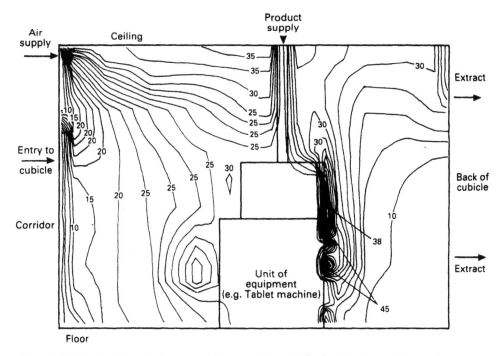

Figure 5.7 Section through the centre of a processing cubicle showing temperature contours

premises and equipment in order to overcome these problems. Wherever possible, enclosed dust-containing manufacturing systems should be employed. Separate 'dedicated' facilities may be necessary for manufacture, involving highly potent toxic or sensitising materials.

and

Suitable environmental conditions should be maintained by installation of effective air-extraction systems, with discharge points sited to avoid contamination of other products and processes. Filtration or other systems should be installed to retain dust. The plant should be properly serviced and maintained. Care should be taken to contain any dust loosened when filters are removed or replaced and to avoid dust falling back on to product from extraction duct-work. The use of tablet and capsule de-dusting devices is recommended.

Dust particles that create most problems lie in the 0.2 to 5.0 μm size range, and collection systems to handle these particles require specialist knowledge when designing the system. Compared to liquids where an extensive library of data is available on their classification into flammable and non-flammable solvents, solids present a more formidable problem. There are extensive data available on a wide range of excipients such as starch, lactose, and sugar, but very little on many of the 'house speciality' chemicals–drugs that have evolved from the in-house research laboratory. This is particularly true in the early stages of development when very little of the compound is available. Apart from its toxicity, any fine particulate

59

material requires three prerequisites for a dust explosion:

- a combustible dust
- a gas which supports combustion
- a source of ignition.

Each system must be fitted with explosion venting of some type specific to the requirements of the materials being processed. In the installation of a dust collection system a number of factors must be addressed:

- The dust collection system must keep the area free of particles. This will depend on the classification of the area, i.e. Class 100 000.

It will require detailed knowledge of the process and whether localized extraction points are required. Close liaison between the process engineer and the dust control engineer will be needed to evaluate and optimize the system.

- The system must contain all material collected.

This means that the filter media and pore size must be carefully selected to ensure that all fine particles are trapped. It may be that a two stage filtration system is necessary if the particle size range is very wide. In this case the secondary filter would retain all particles above say 50 μm and the second finer filter would remove the remainder. HEPA filters may be needed to ensure complete entrapment.

With the trend toward more integrated process manufacturing systems in-house centralized dust collection has become more common, resulting in a reduction of the number of units of equipment required in the processing area. This is in line with the general philosophy of GMP in removing all process utility services to a lower grade technical location. It should be noted that where large volumes of air are being removed from an area there must be an interface with the HVAC. Even where stand-alone dust extractors are used, the heat gain (Figure 5.7) and air volumes required need to be built into the HVAC system design calculation.

The size and cost of a dust control system is directly related to the volume of the entrained air. It follows, therefore, that the correct assessment of this volume is the fundamental prerequisite for achieving both efficient and economic dust control. A typical in-house centralized dust collection system is shown in Figure 5.8.

Compressed Air

All pharmaceutical facilities need compressed air for the operation of equipment, the use of vacuum cleaning systems, spray systems, breathing air, and as instrument air in hazardous areas. It has taken many years to establish the need for clean air for all these operations. The results have been impressive:

- significant reduction in contamination
- reduced maintenance requirements
- less energy requirements
- improved productivity.

Generally, air is supplied through a filtration and compressor unit which draws on

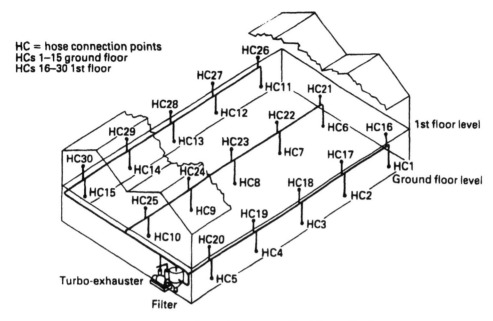

Figure 5.8 Isometric representation of an in-house centralized dust collection system

atmospheric air and compresses to about one eighth of its volume. This air will contain oils, carbons, yeasts, bacteria, dusts (depending on the pore size of the primary filter), and water vapour. It should be remembered that in many operations the compressed air will be in product contact at some point in the manufacturing operation, whether this is in processing, i.e. spray coating operation, or packaging. This will require validation to standards demanded by the state regulatory bodies such as the FDA in the USA, the MCA in the UK, and the European Community. The standards for the quality of air generally required are:

Hydrocarbons: not more than (NMT) 5 mg/litre
Particles: NMT 100/m^3 of 5 μm or larger
Microbial content: NMT 5 colony forming units per 120 litres of air,

and a moisture content of less than 1.0%. In a pharmaceutical facility the systems components will generally comprise a compressor(s), a 1.0 μm filter, an air receiver (sized depending on the capacity requirements of the system; air atomizers for spray coating require large volumes of air), an air dryer (micronizers-fluid energy mills – require large volumes of dry, oil free air), a 0.01 μm filter and distribution piping with point of use valves and filters. The materials of construction used in the distribution are usually 316L stainless steel which reduces the level of corrosion at the product contact outlets. Valves and ball valves have stainless steel balls with PTFE locating seats.

References

EGOZY, Y. and DICK, E. M. (1987) High Purity Water by Electrodeionisation: Principle and Performance, 5th Annual Membrane Technology Planning Conference, 21–23 October, Cambridge, MA, USA.

GANZI, G. C. (1989) The Ionpure Continuous Deionisation Process: Effect of Electrical Current Distribution on Performance, 80th Annual AIChE meeting, 28 November, Washington, DC, USA.

GANZI, G. C. and PARISE, P. L. (1990) The production of pharmaceutical grades of water using continuous deionisation post reverse osmosis, *Parenteral Science and Technology* **44**(4), 231–41.

GANZI, G. C., EGOZY, Y. and GIUFFRIDA, A. J. (1987) High purity water by electrodeionisation, *Ultapure Water* **4**(3), 43–50.

GUIFFRIDA, A. J. (1990) Electrodeionisation Method and Apparatus, US Patent No. 4,925, 541, 15 May.

United States Pharmacopoeia, 21st revision.

6

Considerations in the Design of a Pharmaceutical Facility

MICHAEL P. WALDEN AND GRAHAM COLE

Regulations and Design Standards

'Would you tell me please what that means' said Alice

Like Alice it would be useful to understand the rules before the design is started: 'Pencil erasers are cheaper than concrete erasers'.

Manufacture of pharmaceuticals is a highly regulated business. Quoting John Sharp from the introduction to the third edition of the UK Department of Health *Guide to Good Pharmaceutical Manufacturing Practice* published in 1983: 'The object ... is initially the assurance of the quality of the product and ultimately the safety, well-being and protection of the patient'. It is now almost 35 years since the first legislation, US Federal Standard 209, was introduced to try and regulate the developing contamination control industry. That was in 1963, and it remains today the principal standard used throughout the world. It was followed by the British Standard BS 5295:1976 and similar legislation in other parts of the world. The early standards were quite confusing documents combining mandatory requirements with advice on good facility design. They have been misapplied to give credibility to a range of associated products as diverse as wall finishes, pens and paper. Regulation is now undertaken by agencies representing the major markets into which the finished pharmaceuticals are to be sold, the principal ones being the US Food and Drugs Administration (FDA), the UK Medicines Control Agency (MCA) and now the European Medicines Evaluation Agency (MEA). The process of regulation is effected by a system of product licensing which requires a thorough evaluation and testing of a new drug through a clinical trials programme and then its manufacture, packaging and distribution in approved facilities, using specified equipment and processes and appropriately trained staff. These manufacturing operations use a system of operation known as Good Manufacturing Practice (GMP), sometimes referred to as current Good Manufacturing Practice (cGMP).

Both principal regulating agencies (FDA and MCA) issue guidelines covering all aspects of the manufacturing process. These include notes on manufactu
ties and are the first area of legislation we shall consider.

They are supported by a number of national and international standards, effectively specifications for particular aspects of the process or construction, and we shall also consider some of these.

Continued compliance with GMP standards has always been monitored by frequent audit by the regulating authorities, normally undertaken annually. The desire to make these audits more effective has resulted in the development of a system of validation for facilities, systems and processes. We have become quite preoccupied with validation in recent years and you will read much about it both in this book and elsewhere.

This chapter attempts to establish the GMP ground rules affecting the facility designer.

Good Manufacturing Practice

So many advertising references are made these days to the design, construction or use of 'GMP facilities' that it would be easy to believe that availability of an unlimited budget or installation of the latest 'state of the art' equipment will guarantee the objective. Nothing, of course, could be further from the truth. Every aspect of manufacture from research and development, formulation, staff training, documentation, and quality assurance to storage and distribution is an equally important factor in the process.

The first UK guide was published in 1971, again quoting John Sharp, 'before any formal inspections of pharmaceutical manufacturers had been carried out under the Medicines Act. It was therefore written at a time when the nature, extent and the special problems of the manufacture of medicinal products in the United Kingdom were not completely known'. The UK guides have always been advisory with specific reference given on interpreting the document. 'Should' rather than 'must' is the normal terminology with instructions such as 'ensure that' or 'eliminate' always to be interpreted as implying 'as far as is reasonably practical'. Not surprisingly therefore the advice given has always been very generalized and a great deal of experience and consultation has been required by design professionals (and pharmacists and chemists, production staff and teachers) to ensure that appropriate decisions have been made.

The superseding of the *Orange Guide* by its European equivalent document has introduced one significant change. Since 1 January 1992 compliance with the Guide has been mandatory. In this respect, Europe has now come much more closely into line with the USA where the FDA has always had enforcement powers. As more and more companies look for opportunities to sell their products in the American market or American companies seek multi-site manufacturing opportunities, so the GMP requirements of the FDA and its enforcement policy have become a more serious issue in Europe.

GMP and therefore process validation has traditionally been applied to secondary pharmaceutical production. Raw materials for secondary processing have, however, always been quarantined on receipt while compliance with specification and quality checks have been completed. It was inevitable in this climate that ultimately the regulators would turn their attention to other steps in the production process, primary fine chemicals, biotechnology processes and also the production of containers and packaging materials. So this is the key factor to be considered!

If patient protection is to be safeguarded by a system of good manufacturing procedures and these are enforceable in law, then there must be a system for monitoring compliance. The sheer scale and diversity of the industry makes it impossible for those regulating it to provide definitive requirements relevant to all applications and so the GMP guidelines, both FDA and European, tend to be a collection of general principles rather than hard and fast rulings and monitoring compliance is effected by the validation programme.

The paragraphs relating to facilities in the US guide to GMP are contained in the *Code of Federal Regulations*, Section 21, paragraphs 211.42–211.58, published by the FDA. Facilities meeting FDA standards will generally be acceptable to the MCA but the reverse is not true. Cross approval of facilities is long overdue but still some way away. This short section of 17 clauses contains all the guidance offered on facility design. It frequently uses words like 'adequate', 'appropriate' and 'suitable' without being more specific. Typical is regulation 211.42(a) which states: 'Any building or buildings used in the manufacture, processing, packaging or holding of a drug product shall be of a suitable size, construction and location to facilitate cleaning, maintenance and proper operation'. It is not considered practical to offer more specific guidance as each application associated with every type of process will vary and so the regulators have restricted themselves to a set of general principles. Interpretation of these principles and their incorporation into the design of facilities therefore requires considerable experience on the part of the designers. Regulators remain ready to offer opinions on proposals but are not generally willing to offer direction.

Air Cleanliness Standards

For many years Federal Standard 209E and British Standard BS 5295:1989 have been the principal standards used for regulating environmental cleanliness, clean air. Compliance with these standards has been an essential part of meeting GMP requirements. Although it was possible to cross reference the two, there are subtle differences between them concerning implementation which has necessitated compliance with both. In May 1996 the draft for the proposed international standard ISO 14644-1, 'Classification of Airborne Particulate Cleanliness in Cleanrooms and Clean Zones' was issued for discussion. As of 1 October 1997 British Standards Institution says it will now become a formal standard in September 1998. Convergence of views is important for all concerned with international business but it is also important to understand the hierarchy of implementation for national standards. This explains why it has not been possible in the United Kingdom to discontinue use of BS 5295 in favour of the more widely accepted Federal Standard 209E.

In the UK there has been for many years a standards agency, the British Standards Institution (BSI), which was until recently a government agency. Wherever British Standards exist, compliance is an operational requirement for all government departments, including the Medicines Control Agency. BSI is required to review and update standards five yearly and to rewrite them fully when necessary. Within the European Union the European Committee for Standardization (CEN) has been working to harmonize standards between member countries. CEN standards automatically supersede national standards within the EU and compliance is mandatory for all. Immediately BS 5295:1989 was issued, BSI made application to CEN for work to be started on a European cleanroom standard and a

Technical Committee TC243 was established to oversee this work. A draft standard has been available for several years. Without going into too much detail, the new standard appeared to drive up cleanliness standards and by implication construction costs because of the way its class limits were structured. The CEN Standard has never been formally introduced and has now been superseded by a commitment to produce a truly international standard acceptable both in the USA and in Europe.

Convened in 1992 by the International Organization for Standards (ISO), a Technical Committee (TC209) and a number of working groups are currently well advanced with their deliberations. Table 6.1 identifies the documents currently in preparation and the dates by which it is anticipated drafts will be available for discussion. Formal issue normally follows in approximately nine months. From this table it can be seen that Draft International Standard ISO-14644-1, 'Classification of Airborne Particulate Cleanliness for Cleanrooms and Clean Zones' is the only section currently available. Details of the proposed new cleanliness classes are given in Table 6.2. Cleanliness classifications appear to have been harmonized with existing Federal and UK standards and the threat posed by the CEN standard, which will now be abandoned, appears to have disappeared. Table 6.3 provides an illustra-

Table 6.1   ISO/TC209 – Cleanrooms and associated controlled environments

	Draft international standard available by
Clean air standard	
ISO 14644-1 Classes of Air Cleanliness	May 1996
ISO 14644-2 Cleanroom Monitoring	May 1997
ISO 14644-3 Metrology and Test Methods	October 1997
ISO 14644-4 Cleanroom Design and Construction	May 1997
ISO 14644-5 Cleanroom Operations	October 1998
ISO 14644-6 Terms, Definitions and Units	February 1999
ISO 14644-7 Minienvironments and Isolators	February 1999
Biocontamination Standard	
ISO 14698-1 Biocontamination Control General Principles	February 1997
ISO 14698-2 Evaluation and Interpretation of Biocontamination Data	February 1997
ISO 14698-3 Biocontamination Control of Surfaces	February 1997

Table 6.2   Proposed standard ISO 14644-1 – Classes of air cleanliness: airborne particulate cleanliness classes

Class	Number of particles per cubic metre by micrometre size					
	$0.1~\mu m$	$0.2~\mu m$	$0.3~\mu m$	$0.5~\mu m$	$1~\mu m$	$5~\mu m$
ISO-1	10	2				
ISO-2	100	24	10	4		
ISO-3	1000	237	102	35	8	
ISO-4	10 000	2370	1020	352	83	
ISO-5	100 000	23 700	10 200	3520	832	29
ISO-6	1 000 000	237 000	102 000	35 200	8320	293
ISO-7				352 000	83 200	2930
ISO-8				3 520 000	832 000	29 300
ISO-9				35 200 000	8 320 000	293 000

Table 6.3 Comparative analysis of past, current and proposed particulate standards

US Federal Standards 209E		British Standards 5295		EC GMP Guide	ISO Standards 14644-1	0.5 Micron particles/M^3
English	SI units	1989	1976			
					ISO-1	0*
					ISO-2	4
	M1					10
1	M1.5	C			ISO-3	35
	M2					100
10	M2.5	D			ISO-4	352
	M3					1000
100	M3.5	E–F	1	A–B	ISO-5	3520
	M4					10 000
1 000	M4.5	G–H			ISO-6	35 200
	M5					100 000
10 000	M5.5	J	2	C	ISO-7	352 000
	M6					1 000 000
100 000	M6.5	K	3	D	ISO-8	3 520 000
	M7					10 000 000
		L	4		ISO-9	35 200 000
		M**				(350 000 000)**

* Reference value; counts to be conducted at smaller particle sizes for statistical accuracy
** Reference value; added by author to place class levels in context – these values do *not* appear in the standard. Minimum levels specified are Class M = 450 000 @ 10 microns. Applies to BS 5295:1989 only

tion of how the new standard will sit alongside other standards currently in every-day use.

The general principle of ISO standards is that compliance is at the discretion of the participating countries. It should be noted, however, that within the EU, ISO standards automatically supersede CEN standards and compliance is therefore mandatory.

As standards are updated, there is left behind a significant number of facilities built to standards which are now obsolete. In the past it has been the practice to allow these facilities to continue in production, tested in accordance with the standards to which they were built, so long as the reference standard is clearly acknow-ledged, i.e. 'Class 2, BS 5295:1976'.

Without sight of the full ISO document it is not possible to state with confidence that this will continue to be the case but it is a reasonable assumption. As existing standards may remain in use for some time they will now be considered in some detail.

British Standard 5295:1989

In addition to merely updating the original 1976 standard, BS 5295:1989 introduced a number of important features. It separated the standard into five distinct parts:

- Part 0 – Definitions.

- Part 1 – Specification for clean rooms and clean air devices – including referenced test methods for testing and certifying compliance.

- Part 2 – Method of specifying the design, construction and commissioning of clean rooms and clean air devices.

- Part 3 – Guide to operational procedures and disciplines applicable to clean rooms and clean air devices.

- Part 4 – Specification for monitoring clean rooms and clean air devices to prove continued compliance with BS 5295:1989 part 1.

Parts 1 and 4 are mandatory, and parts 2 and 3 give guidance; separating the mixture of advice and requirement has reduced the confusion which marred the earlier standard.

The revised standard:

- classified cleanliness levels by letter and significantly increased the number of classification levels

- introduced a mandatory requirement to demonstrate continued compliance at specified intervals.

BS 5295:1989 is the standard used by the UK Medicines Control Agency (MCA) in evaluating contamination levels within manufacturing areas. However, a significant number of pharmaceutical companies are either multinational with a US parent company or intend to sell their products within the USA. In such cases standards set by the US Food and Drug Administration (FDA) must be complied with which imposes contamination levels established by US Federal Standard 209. Revision E of this standard was issued in 1992.

US Federal Standard 209E

Generally accepted throughout the world as the premier contamination control standard, revision E contains a number of changes which reflect both its growing importance as an international standard and the increasing sophistication of the cleanroom industry. Among the many changes the following are key:

- Use of metric units as the primary definition of quantitative measure; English units will be used in parallel. The metric (SI) class limit will be predicated on the base 10 logarithm of the number of particles 0.5 micron and larger, per cubic metre of air; e.g. in SI M 3.5 is equivalent to the current Class 100.

- Extension of class limits both above Class 100 000 and below Class 1.

- Verification measurement to be made at one or more particle sizes. The standard test particles sizes are 0.1 micron, 0.2 micron, 0.3 micron, 0.5 micron and 5.0 micron; however, alternative particles sizes can be used as long as the provisions in the document are followed.

- Inclusion of a requirement that particle concentrations be reported as particles per unit volume of air, regardless of sample size. The overall sample size is still to be reported, and the standard will continue to allow different sample volumes at different locations.

- Of the change proposed for FED-STD-209E, one of the most noticeable for all users is the revised format for expressing classes, which incorporates the particle size or sizes for which class limits are specified. This change will help to further standardize cleanroom specifications and terminology.

Other Standards

While the above mentioned documents will continue to be the principal standards of interest to the facility designer, there are nevertheless a significant number of others where some reference may be required from time to time.

- BS 3928:1969 'Method of Sodium Flame Test for Air Filters'. This standard is the reference document for establishing the efficiency of HEPA filters in the UK.
- DIN 24184 'Type Test of High Efficiency Sub Micron Particulate Air Filters'. This is the equivalent German standard. Both are likely to be replaced by a common European standard in the near future.
- BS 5726:1992 'Specification for Microbiological Safety Cabinets'. This recently revised standard is extremely important for those concerned with work in microbiology laboratories. A great deal of effort has gone into updating the specification and also into requirements for testing the containment capabilities of cabinets.
- Australian Standard AS 2567:1982 'Cytotoxic Drug Safety Cabinets' and AS 2639:1983 'Cytotoxic Drug Safety Cabinets – Installation and Use'. These standards deal with cabinets used for processing and reconstitution of these potent drugs. They still represent the only standards worldwide to address this issue.

Those working on FDA inspected facilities are aware that US Federal Standard 209E is supported by a series of IES (Institute of Environmental Sciences) Recommended Practices of which the most widely used are:

- IES-RP-CC-001-83T 'HEPA Filters'.
- IES-RP-CC-003-87T 'Garments required in Cleanrooms and Controlled Environment Areas'.
- IES-RP-CC-006-84T 'Testing Clean Rooms'.

Other documents such as the *Guideline on Sterile Drug Products produced by Aseptic Processing* are published by the FDA to give specific guidance on interpreting the requirements of the Code of Federal Regulations for this particular group of activities.

For those working in biotechnology, *Categorisation of Biological Agents According to Hazard and Categories of Containment 1995*, published by the Advisory Committee on Dangerous Pathogens (ACDP), and *Guide to Genetically Modified Organisms (Contained Use) Regulations 1992 as Amended 1996*, published by the UK Health and Safety Executive (HSE), are important documents containing guidance on containment classification, again establishing general principles useful to facility designers (Table 6.4). The same agency is also responsible for similar guidance for those working with genetic manipulation in both laboratory and large scale applications. Similar classifications to those identified in the table have been published in the USA (see *Federal Register* by the Department of Health and Human Services Vol. 51 No. 88, 7 May 1986, Notices Part III, 'Guidelines for Research Involving

Table 6.4   Tabular summary of laboratory containment requirements

Containment requirements	Containment levels			
	1	2	3	4
Laboratory site: Isolation	No	No	Partial	Yes
Laboratory: Sealable for fumigation	No	No	Yes	Yes
Ventilation: Inward airflow/negative pressure	Optional	Optional	Yes	Yes
Through safety cabinet	No	Optional	Optional	No
Mechanical: direct	No	No	Optional	No
Mechanical: independent ducting	No	No	Optional	Yes
Airlock:	No	No	Optional	Yes
With shower	No	No	No	Yes
Wash hand basin	Yes	Yes	Yes	Yes
Effluent treatment	No	No	No	Yes
Autoclave site: On site	No	No	No	No
In suite	No	Yes	Yes	No
In lab: free standing	No	No	Optional	No
In lab: double-ended	No	No	No	Yes
Microbiological safety cabinet/enclosure	No	Optional*	Yes	Yes
Class of cabinet/enclosure*	–	Class I	Class I/III	Class III

* Required for clinical microbiological suites
Source: ACDP *Categorisation of Dangerous Pathogens According to Hazard and Categories of Containment*, 1995

Recombinant DNA Molecules)'. The European agency (CEN) also has a significant amount of documentation in an advanced stage of preparation.

Finally, the work undertaken by the UK Pharmaceutical Isolator Group should also be recognized. While not a statutory document, its publication, *Isolators for Pharmaceutical Applications*, provides practical guidelines on the design and use of isolators for aseptic processing of pharmaceuticals. It represents an excellent example of users, manufacturers and other interested parties coming together to self regulate an area of new technology currently outside the formal system. Many of these latter issues are likely to fall under the umbrella of the new ISO standards discussed above, again harmonizing the approach required by facility designers.

This brief summary of standards is supported by a bibliography at the end of the chapter.

Design

> *As she said this, she came upon a neat little house …*

The Food and Drug Administration (FDA) and the Medicines Control Agency (MCA) identify design standards of pharmaceutical facilities in only the most general way. For example:

The Federal Code of Regulations Section 21 parts 211.42 states:

a. Any building or buildings used in the manufacture processing, packing, or holding of a drug product shall be of suitable size, construction and location to facilitate cleaning, maintenance, and proper operations.

b. Any such building shall have adequate space for the orderly placement of equipment and materials to prevent mix-ups between different components, drug product containers, closures, labelling, in-process materials, or drug products, and to prevent contamination. The flow of components, drug product containers, closures, labelling, in-process materials, and drug products through the building or buildings shall be designed to prevent contamination.

c. Operations shall be performed within specifically defined areas of adequate size. There shall be separate or defined areas for the firm's operations to prevent contamination or mix-ups as follows:

1. Receipt, identification, storage, and withholding from use of components, drug product containers, closures, and labelling, pending the appropriate sampling, testing, or examination by the quality control unit before release for manufacturing or packaging;

2. Holding rejected components, drug products containers, closures, and labelling before disposition;

3. Storage of released components, drug product containers, closures and labelling;

4. Storage of in-process materials;

5. Manufacturing and processing operations;

6. Packaging and labelling operations;

7. Quarantine storage before release of drug products;

8. Storage of drug products after release;

9. Control and laboratory operations;

10. Aseptic processing, which includes as appropriate;

    i. Floors, walls and ceilings of smooth, hard surfaces that are easily cleanable;

    ii. Temperature and humidity controls;

    iii. An air supply filtered through high-efficiency particulate air filters under positive pressure, regardless of whether flow is laminar or nonlaminar;

    iv. A system for monitoring environmental conditions;

    v. A system for cleaning and disinfecting the room and equipment to produce aseptic conditions;

    vi. A system for maintaining any equipment used to control the aseptic conditions.

d. Operations relating to the manufacture, processing, and packaging of penicillin shall be performed in facilities separate from those used for other drug products for human use.

Similarly, the *Orange Guide* states as a principle:

Buildings should be located, designed, constructed, adapted and maintained to suit the operations carried out in them. Equipment should be designed, constructed, adapted, located and maintained to suit the processes and products for which it is used. Building construction, and equipment lay-out, should ensure protection of the product from contamination, permit efficient cleaning, and avoid accumulation of dust and dirt.

For the manufacture of sterile products the same guide states:

Sterile products should be manufactured with special care and attention to detail, with the object of eliminating microbial and particulate contamination. Much depends on the skill, training and attitudes of the personnel involved. Even more than with other types of medicinal product, it is not sufficient that the finished product passes the specified tests, and in-process Quality Assurance assumes a singular importance.

The *Rules Governing Medical Products* in the European Community Volume IV states in Chapter 3 that:

> Premises and equipment must be located, designed, constructed, adapted and maintained to suit the operations to be carried out. Their layout and design must aim to minimize the risk of errors and permit effective cleaning and maintenance in order to avoid cross-contamination, build up of dust or dirt and, in general, any adverse effect on the quality of products.

This provides architects and facility planners considerable latitude when developing space requirements, layouts, finishes, and details of the facilities. In general however, standards have evolved over the years which means that among user groups in the industry standards are much the same.

The challenge for the architect in the design of a new 'state of the art' facility is that there are many highly sophisticated engineering systems, and the design must incorporate knowledge of highly technical operations such as clean rooms, sterile suites, the integration of complex control systems for the energy management of the building, the process utility services, and the process operation itself with all its associated activities.

A full understanding of the operations to take place within the administration, process area, and flow of personnel and materials through the facility is imperative. The definition of the operation or operations must come from the user groups in the form of written descriptions or preferably logic diagrams – a step-by-step description of operations from one area to another. Gowning procedures and one-way versus two-way directional flows also must be defined. The relationships between the spaces are developed directly from the description of the operations. The spatial requirements and relationships should be identified with input from the user. Often, total space requirements are established before the architects and engineers are commissioned for the project, in which case the architect should evaluate the space allocations, particularly where equipment is to be located. Whether for clinical manufacturing of batches or full scale manufacturing of marketable products, the definition of equipment including equipment sizes, services, and clearances must be established. In addition to space for the equipment itself, adequate space should be provided for items such as carts and/or pallets, control panels, sinks, laboratory benches, balances, and adjacent support space for related equipment and/or services. This last item is one most often overlooked by the facilities planners.

Such support spaces may include service areas for a built-in solid dosage manufacturing items of equipment such as an in-house vacuum cleaning system; autoclaves, ovens or lyophilizers; chased walls for concealing piping or low return ductwork; or service corridors or closets for access to controls, valves and piping.

Once the sizing of primary and support spaces for the facility and process areas has been established, room classifications should be defined. These classifications are in terms of the cGMP level of cleanliness (e.g. Class 10 000). From this information, developed jointly by the designer and client, flow diagrams are developed to establish the organization of the facility, suite, or room(s). The engineers can then use these diagrams to establish room pressurization requirements. It is important to define these parameters at an early stage in order to identify potential problems with regard to the pressurization and provide design solutions (e.g. adding an airlock, subdividing rooms) before final layouts are developed.

The next step is to develop layouts. Even if only a single room is being considered, the layout of the entire area surrounding or serving that room operation must be evaluated in terms of space. The layouts are a translation of the previously developed flow diagrams, incorporating requirements of the operation as well as integrating the anticipated mechanical systems required to service the space or spaces. Full integration of the architectural and engineering designs must take place from initial development of these layouts in order to ensure that mechanical systems will fit, operate efficiently, be easily maintained, and be cost effective. The architect does not have the luxury he or she might have on other types of projects of providing a fixed amount of space above the ceilings and then working independently of the engineers below the ceilings. In pharmaceutical spaces engineering systems are present throughout – above the ceilings, below the ceilings, running horizontally, running vertically, exposed, and concealed. It cannot be too strongly emphasized that the architect must liaise closely with the engineer throughout so that the final architectural space provided is also a well-engineered space. While layouts are being developed, actual wall thicknesses rather than the usual 'single line' delineations must be shown, even at the conceptual design stage in order to assure that properly sized chase walls will be accommodated. For clean room environments 'single-line' layouts are deceiving in that they misrepresent the final gross square footage required for the clean room or suite. Since budgets are often established after the conceptual layouts have been developed, a misrepresentation of total gross square footage can often result in an understating of budget costs for the construction or renovation project.

In a solid dosage form manufacturing facility there are many options available for the process route:

- Direct compression
- Dry granulation
- Wet granulation
- Fluid bed granulation
- Capsule filling
- Powder and granule formulations.

Within these routes there are many variations for the types of equipment used and the sub-routes for the coating processes where a coated tablet is manufactured:

- Sugar coating
- Film coating (solvent and aqueous)
- Enteric coated.

The finishes will need to be resistant to the materials used in this part of the process.

There are also many different approaches to producing sustained release products such as the osmotically driven release pattern. Here two options have been selected and a number of variations discussed.

The conceptual process has been divided into two streams:

- The traditional approach
- The automated material handling approach.

The selected manufacturing solution will depend on the processing philosophy and inventory level control. If operating on a campaign basis, then the facility can incorporate two granulation trains. Alternatively, if more than one wet granulated product is required simultaneously, then the facility will require additional granulating, drying, and milling equipment.

The direct compression tablet lends itself to the dedicated plant concept. A totally enclosed system could be designed to operate with yields of over 95%. It would use a minimum amount of floor space, and reduce expensive room finishes to a minimum. It would take full account of current good manufacturing practice and provide a system that can be validated to the most stringent FDA and MCA requirements, requiring a minimum of analytical support. The degree to which this process can be scaled-up will depend on the data available within the manufacturing company. Certain aspects of the process require careful study by the project team:

- mixing requirements for excipients and especially the lubricants
- compression characteristics of the tablet
- coating system used, e.g. side vented coating pan or fluid bed column (Wurster type)
- aqueous or organic film coating or special processes, e.g. sustained release system
- a detailed study of batch sizes to size the equipment to provide the greatest degree of flexibility where the market requirements change and require alternative products
- powder filled capsules are similar to the direct compression process.

The process for manufacturing could be developed along the same lines. However, as the products require wet granulating, then additional equipment is required for granulation, sizing, and drying, and it would be necessary to operate the plant on a campaign basis. This reduces its flexibility, and the alternative is to use a 'Lhoest' type concept similar to the old SKF facility in Milan. This poses interesting spatial problems for the architect.

All of these processes can be automated, critical control parameters monitored, and essential analytical data generated, using on-line systems to ensure that full documentation is provided for each batch of product. Here the architect is concerned with the layout concepts, people flow requirements, and segregation of different grades of cleanliness.

The design for the manufacture and packaging of various liquid and solid dosage formulations will require all or some of the following:

- Storage for raw materials and active ingredients prior to dispensing.
- Storage of packaging components and labels.
- In-process storage of material during manufacture.
- Storage of packed finished goods (Home Office regulations may apply for a controlled substance for all these stages).
- Dispensing area (this could be used also for sterile operations).

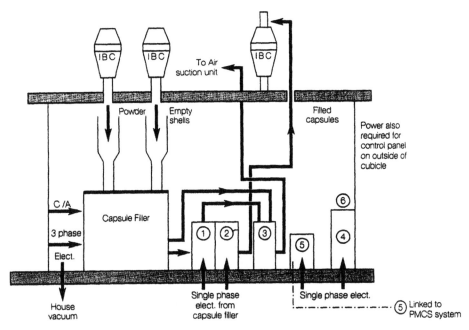

1) Capsule deduster and sorter ′   C/A: Compressed air
2) Metal detector   ELECT: Electricity
3) Interceptor   PMCS: Process monitoring and
4) Checkweigher & bench   control system
5) Label printer   IBC: Intermediate bulk container
6) Scales bench mounted

Figure 6.1   Capsule filling cubicle

- Area to accommodate the equipment. This is physically separate from the dispensary and may have its own technical area for process services. Headroom may be an issue.
- Possible area for blending IBCs.
- Compression or filling area.
- Sterile preparation and filling area.
- Wash area.
- Packaging area.
- Additional packaging area for future expansion.
- Office.
- Male/female or unisex change facilities with shower.
- Engineering workshop.
- QA laboratory.

Equipment

Generally the design philosophy for solid dosages is to have a totally enclosed system from dispensing through to compression or filling and onto packaging, if possible.

75

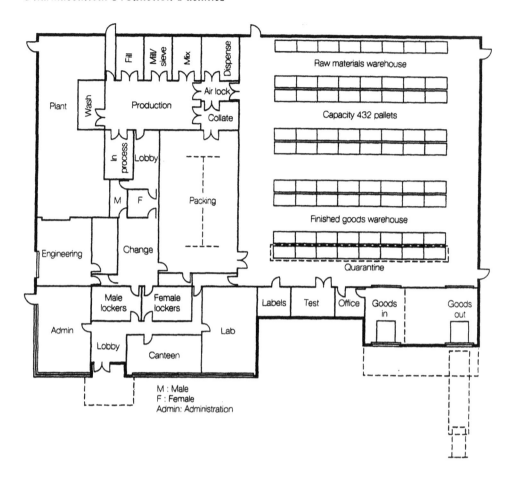

Figure 6.2   Capsule filling facility layout

Excipients will be dispensed into a sack tipper unit, passed through a sieve (800–900 microns) into the mixer–granulator–drier (MGD) using vacuum transfer and stainless steel pipework. The active ingredient is dispensed separately and either added as an alcoholic aqueous solution or dry powder depending on the formulation. The dry mixing part of the process takes place in the MGD. The product is then discharged through a mill into a container. For most capsule processes, granulation is not necessary. At this stage lubricant is added and the container rotated to complete the blending operation.

The container is positioned above the tablet press or capsule filler. The concept is illustrated in Figure 6.1. Tablets or filled capsules are passed through a metal detector and deduster into a container and held until released by QA for packaging.

A typical layout for a capsule filling process is shown in Figure 6.2.

Dispensing

'*I should like to buy an egg please,*' *she said timidly, '*How do you sell them ....*'*

This section discusses the special requirements that are essential in the handling and dispensing of raw materials for the manufacture of pharmaceutical dosage forms. In the pharmaceutical industry there is no real difference in the physical control of the dispensing operation for primary and secondary manufacture. The difference is largely in how these materials are handled and the quantities recorded. Usually primary manufacture handles large quantities of solvent which can be piped to a reactor etc. from a tank farm whereas in secondary manufacture materials, mostly solids, are transferred by a less automated means to a dispensing area adjacent to the manufacturing area.

Here we will examine the criteria that are necessary to design and operate such an area.

The weighing of any pharmaceutical formulation, or, as it is typically called, the dispensing operation, is one of the most critical steps in the production process. All formulae must be strictly adhered to for the required batch-to-batch consistency. Moreover, only approved materials can be used in the operation. Here, as in many other industries, rigid controls must be exercised in order to meet regulatory requirements.

A manual method for weighing materials is a time consuming operation. Each ingredient in a batch formula is carefully measured out, weighed, then checked against the formula for accuracy. At this point, the documentation would be rechecked by one or more check weighers, but even with the tedious checking and rechecking procedure, human errors could still occur. A typical flow diagram which illustrates how material arrives at a plant and is then handled is shown in Figure 6.3.

It is not the purpose here to discuss the procedures that are required to handle and control materials between the supplier, the arrival at the plant and how these materials are tested and released into production. The essential requirement is that material released into the dispensary has been tested and approved by quality control (QC) for production purposes. It is critical that this material is checked by the dispensing operators and that only this material is used for manufacturing products. (In a multi-purpose facility some of the items may be required for other purposes, i.e. research and development, analytical test purposes. It is important to ensure that these materials, which may not have been tested to the same extent or manufactured to the same specification as production material, are not used in the company's products.)

The dispensary handles several different classes of material. There is the bought-in product (the largest range of different materials supplied by approved vendors), and in-house manufactured material similar to intermediates in primary production and products such as Purified Water BP. In addition there are the active ingredients which can be highly potent, cause allergic reactions and also have mutagenic etc. properties.

Material may also have to be passed back to the warehouse when only a partial quantity is required from a large container.

When designing a dispensary the following design attributes should be considered:

- Containment and separation of product from operator.
- Cross contamination.
- Facilities for cleaning booth and equipment.
- Gowning and washing facilities for the operators.

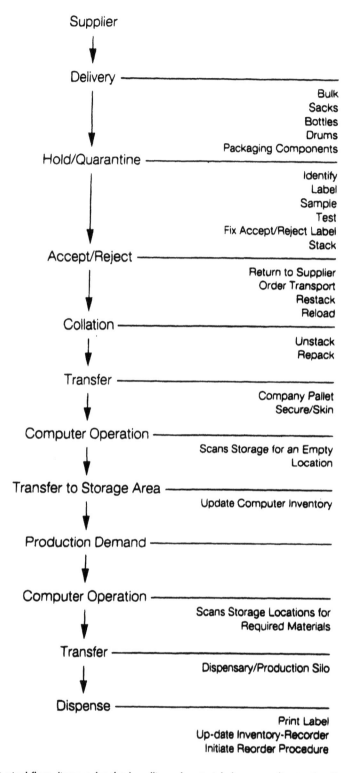

Figure 6.3 Typical flow diagram for the handling of materials from supplier to the dispensary

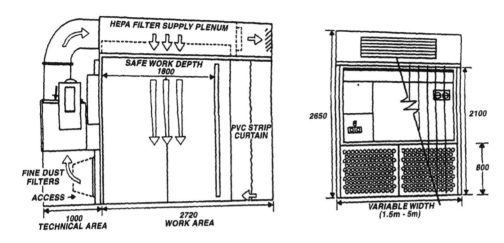

All dimensions in mm

Figure 6.4 Powder handling booth

Figure 6.5 Dispensing booth

- Separation from other areas, e.g. warehouse, corridors, manufacturing.
- What ancillary equipment is required, i.e. scoops, lances, etc.

Dispensary Design

The design of typical powder handling and dispensing booths are illustrated in Figures 6.4 and 6.5. The layout of an area containing in addition washing, gowning and showers is shown in Figure 6.6.

These types of booths are produced by several companies. The one illustrated is manufactured as a stand-alone unit by Extract Technology and is used extensively

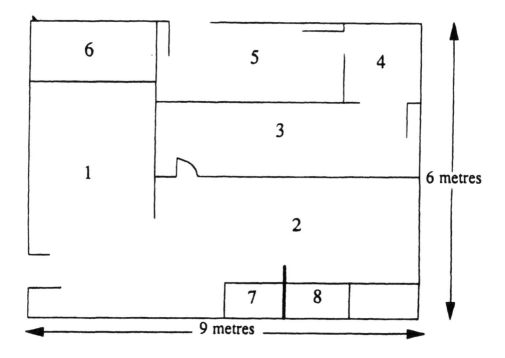

Legend:

1.	Powder Containment Booth	2.	Operation's area
3.	Suit Wash and Storage	4.	Body Shower
5.	Entry-Change	6.	Booth Motors
7.	Sink	8.	Glove Boxes

Figure 6.6 Layout of a complete dispensary

in many industries to combat the dangers of airborne dust in powder dispensing and the sub-division of materials.

The specially designed recirculatory airflow system with high quality filtration provides a Class 100 clean environment and guarantees operator protection while eliminating problems of cross contamination.

The standard dispensing booth is in a modular form which allows for rapid installation and integration into an existing facility. Generally they are constructed in single skin 304 quality 240 grit stainless steel, with external infill wall panels of epoxy coated mild steel. The ceiling plenum and associated housings in the technical area are also white epoxy coated.

The booth's recirculatory airflow incorporates two stages of filtration at low level; fine dust filters followed by mid-stage HEPAs with an airflow distribution screen mounted below the lights in the ceiling plenum. Access to the HEPA filters is from the rear of the booth. Working in a clean environment the operative can generally dispense the product at the 500 mm wide stainless steel bench that is situated in the left hand side of the booth.

An integral frequency inverter provides the smooth running of the motor and allows the fan speed to be adjusted to suit the filter condition. The electrical control panel located behind a removal panel in the rear bulkhead is connected to the motor, lights, power sockets and stop/start stations so that only electrical supply 3 phase, neutral and earth to the panel to enable start-up is required.

The basis of the design will be to comply with Federal Standard 209E. More detailed regulatory issues were covered earlier in this chapter. For the sake of clarity some basic information is provided here.

Cleanroom Legislation (Applicable to Directional Airflow Devices)

Just as in full size cleanrooms, clean air directional airflow devices must comply with the relevant clean room classification. Here we are considering compliance with US Federal Standard 209E. Hence any directional airflow equipment with the operating objective of maintaining a 'Class 100' environment must be clearly validatable to the standard. Unlike cleanrooms which feature large controlled spaces perhaps with airlocks and pressure gradients, the smaller localized clean air environment provided by the directional airflow bench or hood (horizontal or vertical) must be challenged to show the limits of the clean environment envelope. A simple method is to use a smoke generator to establish the boundary of the clean zone, i.e. the specific points at which the directional airflow decays and causes turbulence. The turbulence will entrain contaminated room air, effectively ending the clean envelopes classification. So although the legislation in terms of US Federal Standard 209E is effectively the same, the amount of work required to verify the classification of the smaller clean area of a directional airflow hood is substantially more demanding.

Current Good Manufacturing Practice (cGMP) Compliance (for Directional Airflow Devices)

The area of cGMP compliance covers many aspects of the devices specification. One of the most potentially serious areas of concern to a visiting regulatory inspector is the risk of cross contamination. There are three different possibilities:

Table 6.5 OEL based categorization

Containment category	1	2	3
Compound OEL $\mu g/M^3$	>100	1–100	<1

- Airborne cross contamination
- Structural cross contamination
- Operator borne cross contamination.

Categorization of Compounds

Most of the theory of compound categorization was developed in the USA. Companies such as Eli Lilly and Syntex who pioneered the application of differing containment devices realized that categorizing their compounds into three or four simple groups made the engineers and pharmacists focus on the relevant technology rather than an 'over engineered' solution. As pharmaceutical compounds all have differing characteristics, the operator exposure level (OEL) based categorization in Table 6.5 outlines a simple categorization.

Some compounds are dermal or respiratory sensitizers which means that certain workers will be more susceptible to an adverse reaction. Other compounds may have a cumulative effect. The examples given in Table 6.6, therefore, are guidelines.

Principles of Bulk Dispensing

The modern dispensary facility in a pharmaceutical plant must be capable of:

- Dispensing raw materials to a registered formulation that will be manufactured into a product for either pharmacy or hospital use. The exception to this is a

Table 6.6 Categorization guidelines

Product type	OTC Drugs	Analgesics antibiotics	Immuno suppressants hormonal materials
Typical OEL range Standard	Above 100 $\mu g/M^3$ Category 1 IBC	1–100 $\mu g/M^3$ Category 2 IBC	Below 1 $\mu g/M^3$ Category 3 High containment IBC
Containment devices	Local exhaust hood Bag tip hoods	Downflow powder control booth Drum and cone handling systems	Glove box isolator Alpha/beta port transfer with drum or cassette

IBC: Intermediate bulk container
M^3: Cubic metres
OTC: Over-the-counter

dispensary for research and development, or clinical trials of a product where other considerations may apply.

- Providing records that enable a complete audit trail of all the materials dispensed. This must include both pharmaceutical materials and packaging components. However, packaging components are usually dispensed and reconciled at the beginning and end of a packaging operation in a separate area. Here we will not consider these aspects further.

In today's modern plant there will probably be a 'stock control' and/or a warehouse management system in place. These are also manufacturing facilities which use 'factory management systems', sometimes referred to as MRP systems (material resource planning).

At all times regulations under the Health and Safety Executive, MCA and FDA (if applicable) must be strictly followed. In addition all rules pertaining to Good Manufacturing Practice (GMP) must be obeyed. Standard operating procedures (SOPs) are subject to the 'validation procedure' and all documentation relating to this must be kept secure.

General Hardware Considerations for Dispensing Operation

Balances/Scales

There are many companies which supply suitable balances/scales for a dispensing operation. Three of the most used are: Mettler-Toledo, Sartorius and Avery-Berkel. Normally a dispensing booth will have two or three weighing units. Typically they could comprise:

- An analytical type balance, capacity 1.0 kg to an accuracy of 0.01 mg. This would be used for the active component or ingredients with very low concentrations such as colours.
- An industry standard 0–10 kg scale with accuracy to 500 mg.
- Balance 30–50 kg, accuracy to 1.0 g.
- Within the dispensing area a scale with a much larger capacity is usually situated but is not required in every booth. This is a weighing platform sunk into the floor for weighing up to 1000 kg to an accuracy of 200 g.

A special consideration of all this equipment is the ease of operation, cleaning and maintenance.

Process areas require high quality finishes to maintain cGMP standards. Traditionally pharmaceutical secondary manufacturing facilities have been designed on the basis of single rooms or cubicles for each stage of the manufacturing process. Transfer of materials has been accomplished using a large drum or mobile trolley. Today the industry is investing in more automatic transfer systems for material handling in an attempt to reduce costs and improve yields. This results in a more integrated manufacturing unit. Computer integrated manufacturing (CIM) is becoming more widely used in the pharmaceutical industry to reduce labour costs, improve efficiency and increase yields. It also reduces the size of the building and the high cost areas within that building for each manufacturing operation. The objective

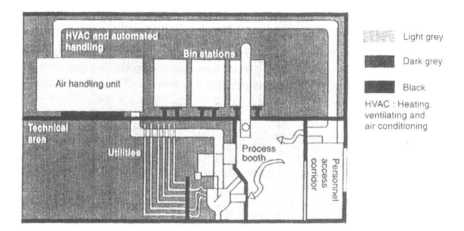

Figure 6.7 Detailed section through an automated manufacturing facility

is to remove as many of the service functions outside the process area into a lower grade technical area. This means that process utilities can be serviced without interfering with manufacturing operation, as illustrated in Figure 6.7. A comparison with Figure 6.8 shows how the expensive process space has been reduced. Figure 6.9

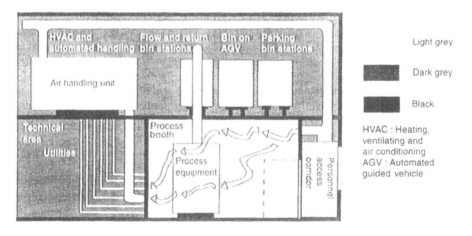

Figure 6.8 Detailed section (contained equipment) through an automated manufacturing facility

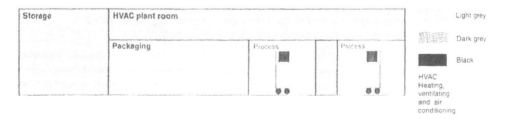

Figure 6.9 Section through a traditional manufacturing facility

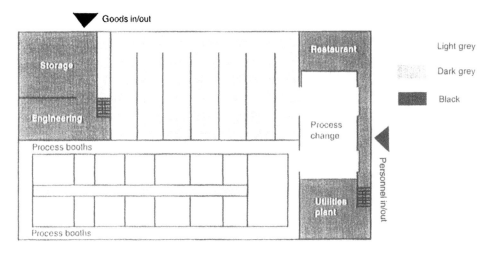

Figure 6.10 Ground floor layout – traditional

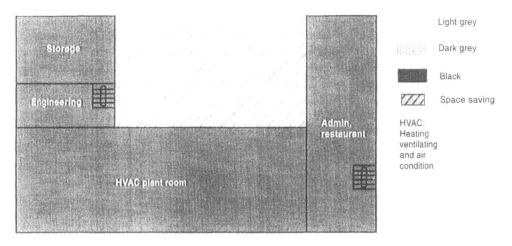

Figure 6.11 First floor layout – traditional

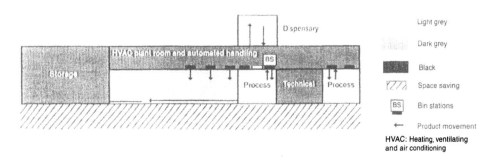

Figure 6.12 Section through automated and non-automated manufacturing facility showing flow of material

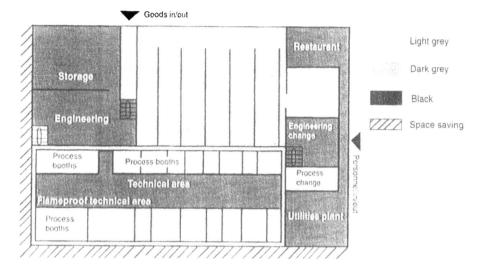

Figure 6.13 Ground floor layout for an automated facility

shows a section through this type of facility. Figures 6.10 and 6.11 show the layout of a traditional solid dose manufacturing facility. All of these drawings have been shaded to show the grading of each area, e.g. light grey, dark grey and black. These gradings refer to the quality of the air required in each area and the quality of the finishes required. Black will require minimum standards and light grey the maximum. Figure 6.12 is a section which shows the flow of material in these types of facility. Figures 6.13 and 6.14 illustrate the concept required for an automated facility. It is important to note here in comparison with Figures 6.10 and 6.11 the

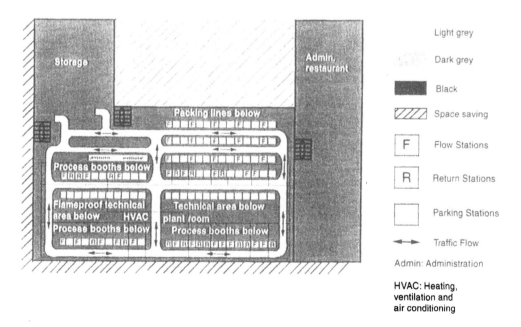

Figure 6.14 First floor layout for an automated facility

86

considerable savings that can be achieved in building costs by reducing the size of the buildings footprint and of the process area. The traffic floor (Figure 6.14) illustrates the pathway for the movement of IBCs for feeding and receiving product from the process equipment on the floor below.

Materials

The requirements as stated in the regulations are very general, and there are many finishes which meet the requirement of being easy to clean, water and chemical resistant, and impervious.

The architect must work with the user groups and facilities planners to identify the finishes most suitable to operational needs within the clean environments and to meet budgetary requirements. Also, the architect's previous experiences with these products, along with manufacturer's information and test data, provide important input into this decision-making process. Finally, the budget considerations should include a review of first cost versus life-cycle cost for the assorted finishes. Some finishes initially may be more expensive than others, but have much longer life spans and consequently may be more beneficial to certain clients in the long run.

The following is a listing of finishes suitable for manufacturing areas:

1. Floors
   a. Epoxy resin – A troweled-on material which is hard, durable, and chemically resistant. Integrated coved bases are available. This is in the medium cost range. It is used in cGMP spaces and related support spaces. While more moderately priced than some of the other floor finishes, it may not be as long lasting as the others.
   b. Welded sheet PVC – Sheet PVC with excellent chemical resistance. Integral coved bases are available. Often used in cGMP aseptic spaces, owing to compatibility with adjacent wall finishes for a fully monolithic environment. It is costly, and if a heavy amount of cart traffic is expected, the welded PVC may be damaged and require maintenance.
   c. Epoxy terrazzo – Provides a hard, durable, long lasting surface. It is the most expensive of the finishes, but its longer life can offset these costs. Granite aggregate chips should be used for best chemical resistance. Care should be taken when using terrazzo to ensure that the filler is not attacked by acid cleaning agents. It is the most durable and works well in high traffic spaces. An integral coved base is available, either troweled or precast, but the troweled base is recommended for the best construction detailing.
2. Walls
   a. Epoxy coating (with filler coat at concrete block walls) – Low cost, durable, and impact resistant. Best for use in areas where high traffic occurs. However, because of the hardness of the material, cracking potential is greater than with PVC coatings and is not as desirable in the aseptic areas.
   b. Polyester coating (with filler coat at concrete block walls) – Similar characteristics to the epoxy coating, with a very durable surface and good for use in high traffic areas though with similar limitations as the epoxy coating for the aseptic spaces.

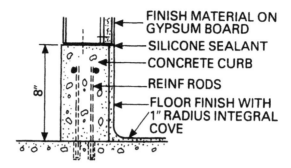

Figure 6.15   Terrazzo base

   c. Seamless PVC coating (with filler coat on concrete block walls) – Sprayed-on coating for use in cGMP areas owing to very good resistance to chemicals and water as well as flexibility to avoid cracking. Not as good with high impact areas, though patching is relatively easy. Quality of installation is most critical for this finish.

   d. Welded sheet PVC – Similar in characteristics to the seamless PVC coating, but because it is thicker than other wall finishes, it is more durable and is the most often used in aseptic areas. It is also the most expensive of the wall finishes.

   e. Prefabricated wall panels – Many variations are available, and each case must be reviewed on its merits regarding material being processed and cost.

3. Ceilings

   a. Gypsum board with:

     Epoxy coating
     Polyester coating
     Seamless PVC coating
     (See comments for these finishes listed under walls).

The architect must also be aware of the finish requirements for outlets and panels that the engineering disciplines will specify for the process utility system.

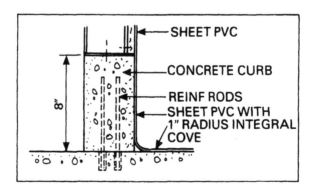

Figure 6.16   PVC sheet base

Generally, such outlets and panels are stainless steel and recessed flush with the adjacent wall or ceiling in order to assure that the finishes are equally smooth, cleanable, water and chemical resistant. Chrome plating of outlets, for example, is not acceptable in many room environments, owing to the types of chemicals used in washing down. Such chemicals cause flaking of the chrome plating.

Details

The architect must translate the generality of the FDA standards into specific details throughout the plant to assure that the 'monolithic' surfaces and finishes provided are maintained. Where openings and/or projections occur within the space, the architect should avoid any horizontal surfaces which could collect dust and promote the growth of bacteria. Although the equipment within the space itself is a dust collector to some extent, it is directly washed as defined in the cleaning protocols.

Generally, in detailed engineering the architects will spend most attention on the penetrations and intersections. Penetrations can include doors, windows, pass-through cabinets or piping, service outlets, air terminals, panels, lighting, and sprinklers. Intersections will include any surface intersection, such as floor to wall, wall to wall (inside and outside corners), wall to ceiling, as well as any joint between dissimilar materials.

For the intersection between floor and wall, the most typical condition is to provide for an integral coved base of the same material as the floor (either trowelled epoxy terrazzo or monolithic sheet PVC) (see Figures 6.15 and 6.16).

References

BS 5726: 1992 'Specification for Microbiological Safety Cabinets, Parts 1, 2, 3 and 4', British Standards Institution.

BS 5295: 1989 'Environmental Cleanliness in Enclosed Spaces, Parts 0, 1, 2, 3 and 4', British Standards Institution.

BS 3928: 1969 'Method of Sodium Flame Test for Air Filters', British Standards Institution.

Categorisation of Biological Agents According to Hazard and Categories of Containment, Advisory Committee on Dangerous Pathogens, HMSO, London, 1995.

Code of Federal Regulations Section 21 Parts 200–299, *US Federal Register*, Washington DC, USA.

Control of Substances Hazardous to Health/Control of Carcinogenic substances – Approved Codes of Practice, Second Edition, Health and Safety Commission, HMSO, London, 1990.

DIN 24184 'Type Test of High-efficiency Sub Micron Particulate Air Filters', Deutches Institut fur Normung V, Berlin, Germany.

DW/142 *Specification for Sheet Metal Ductwork – Low, Medium and High Velocity Air Systems*, Heating and Ventilating Contractors Association, London, 1983.

DW/143 *A Practical Guide to Ductwork Leakage Testing*, Heating and Ventilating Contractors Association, London.

EH40/91 *Occupational Exposure Limits 1991*, Health and Safety Executive, HMSO, London, 1991.

Good Laboratory Practice – The United Kingdom Compliance Programme, Department of Health, London, 1989.

Guidelines for the Large-scale Use of Genetically Manipulated Organisms, Advisory Committee on Genetic Manipulations, ACGM/HSE Note 6.

Guide to Genetically Modified Organisms (Contained Use) Regulations 1992 as Amended 1996, Health and Safety Executive, HMSO, London, 1996.

IES Recommended Practice IES-RP-CC-001-83T *HEPA Filters*, Institute of Environmental Sciences, Mount Prospect, Illinois 60054, USA.

IES Recommended Practice IES-RP-CC-003-87T *Garments Required in Clean Rooms and Controlled Environment Areas*, Institute of Environmental Sciences, Mount Prospect, Illinois 60054, USA.

IES Recommended Practice IES-RP-CC-0056-84T *Testing Clean Rooms*, Institute of Environmental Sciences, Mount Prospect, Illinois 60054, USA.

ISO-14644-1 'Classification of Airborne Particulate Cleanliness in Cleanrooms and Clean Zones', draft for discussion (Eds) International Organisation for Standardisation, Vienna, 1996.

LEE, G. M. and MIDCALF, B. (Eds) *Isolators for Pharmaceutical Applications*, HMSO, London, 1994.

The Rules Governing Medicinal Products in the European Community Volume IV *Guide to Good Manufacturing Practice for Medicinal Products*, Commission of the European Communities, Luxembourg, 1989.

SHARP, J. R. (Ed.) *Guide to Good Pharmaceutical Manufacturing Practice 1983*, Department of Health and Social Security, HMSO, London, 1983 (now superseded by the European Guide 1993).

US Federal Standard 209E *Clean Room and Work Station Requirements, Controlled Environments* US Government Printing Office, Washington DC, USA, 1992.

7

Cleanrooms

MICHAEL P. WALDEN

The door led right into a large kitchen, which was full of smoke from one end to the other.

At the present time the use of cleanroom facilities is widespread in a variety of industries. So diverse are the applications which include microelectronics, defence, food manufacturing, video tape, and compact disc production, in addition to pharmaceuticals and biotechnology processes, that it is often difficult to be specific about the term.

Facilities range from small rooms or suites built within existing buildings to vast purpose-built structures forming entire campuses.

During the Second World War the success of a number of new technological developments, radar and sonar for instance, were restricted by their unreliability in use. Even as late as the Korean war, maintenance costs on many items of equipment exceeded initial capital costs by up to 200% per annum. Many of the failures resulted from key components becoming contaminated at the point of manufacture with minute dust particles contained within the atmosphere of the manufacturing facility. The advent of the space programme with its requirement for smaller and smaller components accentuated the problem. In the pharmaceutical industry also, control of bacterial and pyrogenic contamination was essential if new drugs and devices being developed were to achieve acceptable levels of safety.

The principal function of the cleanroom is the protection of the manufactured product from contamination. In the pharmaceutical industry the lives of patients and the very commercial survival of the manufacturer depend on the integrity of the product. It is important therefore to identify the potential sources of contamination. These may include the working environment, the raw materials, process equipment, and manufacturing personnel.

This chapter is principally concerned with the design and construction of cleanrooms – the means of controlling the working environment. It is, however, important to remember that control of the operating procedures within the room, i.e. the raw materials, process equipment, and manufacturing personnel, is a prerequisite of successful cleanroom operation – good housekeeping. Perhaps the clearest statement of these factors when looking for a definition of the term cleanroom, is found in the IES Recommended Practice, IES RP-CC-006-84T *Testing Clean Rooms* (1984) which states that: 'A cleanroom is a room in which the:

- air supply
- air distribution
- filtration of air supply
- materials of construction and
- operating procedures

are regulated to control airborne particle concentrates to meet appropriate cleanliness levels as defined by Federal Standard 209B or latest issue'.

Raw materials entering the cleanroom can be controlled to a large extent by careful initial selection, preprocessing to clean materials by removing wrappers, etc., and good housekeeping practices as the materials are processed through the room.

Process equipment can be designed to reduce contamination to a minimum by the use of non-shedding materials such as stainless steel. Moving parts such as motors and drives can be encased, or removed from the cleanroom altogether. Services can be installed to provide the minimum interruption of the clean space with a total absence of horizontal surface where particles might settle.

People, manufacturing personnel, remain the single biggest source of contamination in the manufacturing process. The point is well illustrated by the graph in Figure 7.1. It shows the relative contamination levels throughout a working day and the effects of people movements within the room both during production and during evening janitorial activity. Cleanroom standards generally recognize three states in which the room may be validated:

- 'As built', a test principally used to establish that the cleanroom constructor has met his contractual obligations.
- 'At rest', a test taken with the room fully equipped ready for production but with no equipment running and personnel absent.
- 'In use', a test taken with the room fully operational.

From Figure 7.1, it can be readily appreciated that there is a significant difference in the particulate burden in the room between a test conducted 'at rest' at 6.00 am and a test conducted 'in use' at midday or at 8.00 pm when the cleaners are at work.

It becomes essential, therefore, for the cleanroom designer not only to understand the engineering standards and construction techniques required if the cleanliness levels are to be achieved but also to understand how the room will be used and more importantly how the room's performance will be validated before the design is commenced. These issues are addressed in the following paragraphs.

There is one further point by way of introduction. Reference was made earlier to the need to protect the product from contamination by manufacturing personnel. There are an increasing number of instances where, owing either to the toxicity of the product or in biotechnology applications for instance, where there may be manipulation of live organisms and viruses, it is necessary to protect the personnel from the product. Cleanroom technology has developed to provide this protection in a variety of ways. Negative pressure rooms may be used for secondary containment of enclosed manufacturing processes, or manipulative work may be conducted in safety cabinets or flexible film isolators to ensure the right level of protection is afforded.

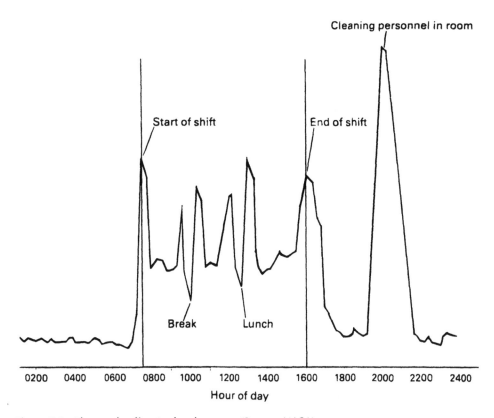

Figure 7.1 The people effect in the cleanroom (Source: NASA)

Although experience is far more limited, comprehensive guidelines on construction standards nevertheless exist in both the USA (1986) and UK (Note 6).

Planning

Cleanrooms within the pharmaceutical facility cannot be considered in isolation. The decision to use designated clean spaces cannot be taken lightly. They are expensive places to build and to operate. Indeed, the term cleanroom can be applied to a wide range of areas from those with minimal environmental and housekeeping controls to aseptic areas using laminar flow air distribution, total change of working garments and the most vigorous cleaning and sterilizing regimes.

The facility design team is faced with balancing a most difficult equation:

Minimum Risk Versus Optimum Cost

In other words, what level of cleanliness is required to minimize the risk of product being contaminated while at the same time producing a facility which is both economical to construct and run without adding prohibitively to the product unit cost. Typically, where product can be terminally sterilized after production by steam, by gas, or by irradiation, the post-sterilization contamination risk may be greater than 1×10^6. However, many products, particularly where there is manipulation or reconstitution, may be totally unsuitable for sterilization and therefore the whole production technique must be enhanced by the use of aseptic rooms to reduce that contamination risk. Even so, risk levels around 1×10^3 would not be unusual. The question of production technique has got to be considered in the context of the total building philosophy. Cleanroom design should not be allowed to progress until the fundamental questions affecting any process building design have been considered. Typically these would include:

- Process flow
- People flow
- Services distribution
- Environmental conditions
- Cleanliness regime
- Exit in emergency
- Relationship with ancillary areas
- Relationship with external facilities, including plot boundaries
- Specific legislative and user standards.

Decisions on building philosophy will impact significantly on cleanroom design but more importantly on the services design, particularly heating, ventilation and air conditioning (HVAC). The key question will be, is the facility to be purpose designed, dedicated to a single product or group of products for the life of the building, or will the 'serviced shed' concept be used? In the former, large centralized HVAC plants may provide all the building's air requirements, and provided that there is no risk of cross-contamination, will do it most efficiently. On the other hand, the 'serviced shed' where production areas will have comparatively short life spans will have a number of smaller, modular plants serving individual rooms or suites of rooms, giving flexibility to operating conditions and adaptability to regular upgrade, perhaps at the cost of economy.

The cleanroom designer, in isolation, will not resolve all the issues. Successful cleanroom design requires a dedicated team approach by process engineers, production personnel, and the quality assurance department. In large scale projects there is a significant possibility that construction personnel and consultants may be unfamiliar with cleanroom construction techniques. Early legislation was very restrictive in the way it influenced design. It often misled the designer rather than clarified performance issues. Whilst revised standards have attempted to tighten up on specifications, some give much clearer guidance on design. British Standard 5295:1989, for instance, in recognizing the importance of room state on particulate levels during testing, also recognizes that the designer will need additional information for each level of testing that the room is required to meet. The Standard devotes an entire section, Part 2 'Method of specifying the design construction and commissioning of cleanrooms and clean air devices' (BS 5295: 1989) to giving guidance on design, and Table 7.1 provides an invaluable check list to the design team on the information necessary to develop the design. The British Standard, in fact, places a mandatory obligation on the customer to provide the information if the designer is to be contracted to meet specific test criteria.

Whilst cleanroom planning can be considered only in general terms, a number of specifics can be addressed. Of particular importance are the ways production personnel and materials access the cleanroom. Invariably an airlock system is used to

Table 7.1 BS 5295: Part 2: 1989 Table Summary of information required. Note: This figure is abbreviated for clarity. The original contains references to appropriate specification clauses

Information required	Relevant clauses	
	Essential information	Additional information required
Reference (i.e. the number and date of this British Standard)	•	
Class of environmental cleanliness and state of occupancy	•	
Purpose of the controlled space	•	
Layout	•	
Dimensions	•	
Services	•	
Responsibility for commissioning	•	
Design		•
Construction		•
Additional sampling positions		•
Monitoring positions		•
Temperature and humidity		•
Lighting and glare		•
Noise and vibration		•
Ventilation and air change rates		•
Microbiological contamination		•
Filtration systems		•

avoid direct access between the cleanroom 'white' areas and the wider facility beyond, often referred to as 'black'.

Change Room Philosophy

Historically, the design and use of cleanroom changing rooms has done little to instil the sense of discipline in staff and operators by which the complete contamination control activity is maximized. Often the return on the very high investment has been eroded by this oversight.

The modern cleanroom requires that management should enforce, and that staff should follow, strict personal habits. This starts in the changing areas where the layout and 'flow' should be progressive from 'black' through 'grey' to 'white' zones.

The 'black' zone is used for changing and storage of outer clothing. Normally equipped with individual lockers, there should be provision for storage of wet and heavy outdoor clothing and footwear as well as a proportion of crash helmets, etc. Flooring should be readily cleanable, and entrances should be guarded by contamination control mats. 'Internal' footwear may be provided at this stage. The 'black' zone change room may be located away from the cleanrooms, close to the employee's entrance to the building, or it may form part of a double change entry procedure.

The 'grey' area is where specialist 'underalls' may be held when applicable. Separate male and female sections can be required if personal domestic clothing is changed for uniform undersuits. Again, a personal, secure locker will be required where a total garment change is implemented. The flow progresses to storage and use of cleanroom clothing proper where the installation will depend on the clothing selection and change frequency. Facilities should be available in the 'grey' area for staff to 'scrub-up' and to remove make-up where necessary. The actual method of dispensing cleanroom garments may vary from vertical laminar flow control cabinets to simple rails with hangers. Flooring should be contamination controlled and lead to the 'white' area.

The 'white' area is where staff change into their cleanroom footwear and 'step-over' into the cleanroom via a bench or swivel-seat divider. Ideally, the complete changing environment would incorporate an air flow moving from 'clean air' at the 'white' zone through 'grey' to the 'black' zone. The changing room must also contain facilities for dispensing cleaned sterile clothing, masks, and gloves and, no less important, equipment for the collection of soiled items to be prepared for processing. Showers may be provided for emergency use or for routine use in areas where there is risk of personnel contamination by toxic substances.

Material Transfer Techniques

Material movements throughout the pharmaceutical process are important, but their introduction to and removal from cleanrooms must be carefully controlled if they are not to introduce contamination.

Wherever practicable, even raw materials must be manufactured and packed in clean conditions. Polythene and similar wrapping materials should be used in preference to paper. Where shipper cartons are necessary they must be removed and the contents decontaminated well before materials enter the cleanroom.

Materials must be transferred through airlocks, and wherever possible dedicated vehicles or trolleys should be used, avoiding the need for one truck to pass from 'grey' to 'white' areas. Pallets, when used, should always be of plastic construction.

Smaller volume materials can be transferred through air lock hatches, using dedicated trays if necessary. Where conveyor systems are used it may be necessary to introduce separate conveyors with dead plate transfers at the cleanroom entry point to avoid transferring contamination from 'grey' area to 'white'.

Even fluid materials may require filtration at the point of use to guarantee the absence of particles once in the manufacturing process.

While great emphasis has been placed in this chapter on particulate contamination, it must be recognized that certain powder based processes such as tabletting and vial filling are used when handling a particulate material. The concern in these situations is to prevent the ingress of foreign particulate matter. It is also important to reduce the risks to personnel and to the environment by explosion, for instance.

The following paragraphs investigate key aspects of construction against the skeleton of the cleanroom definition quoted above.

1. Air Supply

It has already been established that clean air technology is principally concerned with airborne particulate contamination, the number of particles of a specified size measured in a given volume of air at a specified point.

For this reason, the first three aspects of the cleanroom definition – air supply, air distribution, and filtration – are very closely interrelated. Decisions on cleanliness affect filtration requirements; these in turn affect supply air quantities and so on.

Table 7.2 Comparative analysis of cleanroom data

Fed. Std. 209E classes	BS 5295:1989 classes	*No. of 0.5 μm particles/cu./ft	Filter efficiency	Air changes/hr (tested AT REST)	Air changes/hr (tested IN USE)	Min. pressure differentials	
						Classified/ unclassified	Classified/ lower class
	A		For future use				
	B		For future use				
1	C	1	99.9999%	†Unidirectional	†Unidirectional	15	10
10	D	10	99.999%	†Unidirectional	†Unidirectional	15	10
100	E–F	100	99.997%	†Unidirectional	†Unidirectional	15	10
1000	G–H	1000	99.97%	50	60–80	15	10
10 000	J	10 000	99%	25–30	40–50	15	10
100 000	K	100 000	95%	20	25	15	10
	L	Classified only <5 μm particles	85%	10	‡	10	10
	M (Controlled areas)	Classified only <10 μm particles	70%	5	‡	10	—

* British Standard class limits are set in particles/m³ and represent only 86% of US values when converted to ft³. As the majority of particle counting machines measure in ft³ the figures given in this column must be adjusted when validating a British Standard classified facility.

† The majority of unidirectional rooms constructed have a downflow air pattern with terminal filter velocities of 90 ft/min (0.45 m/sec) +/− 20%. In normal applications this equates to 5–600 air changes per hour. Devices may be incorporated into control systems to permit up to 50% decrease in air volumes during AT REST periods, principally as an energy saving arrangement.

‡ Site conditions vary.

In considering these items individually, it should be remembered that decisions made on one aspect will influence the others.

The key question affecting air supply is one of quantity. How much air is necessary in each area? This decision influences the size of air handling plant, the quantities of utilities such as power, steam, and chilled water, and will also determine the size of services zones necessary to accommodate ductwork and plant rooms to accommodate generating equipment.

It is, therefore, essential to consider at the outset of the HVAC design whether or not the quantity of air to be supplied to any room is heat load determined or not. That is, is the air volume necessary to dissipate the heat load within the room significant or not? Engineers trained to calculate their designs with precision are constantly frustrated by the very imprecise nature of cleanroom design. The validation requirements to meet legislated class limits are precise enough, it is true, but there is no formula capable of relating air change rates to cleanliness level. The designer must use his experience to gather precise information about the nature of the operations within the room, the number of people concerned, the nature of the equipment, and the state in which the room is to be tested, i.e. 'at rest' or 'in use'. He must then make a judgement on the air quantity required. Standards have, in the past, specified a minimum number of air changes, without reference to cleanliness level, for instance 20 per hour. Unfortunately, this has caused considerable misunderstanding and has frequently been quoted in contract specifications as an absolute requirement. Table 7.2 is, therefore, offered in the full knowledge that it is also likely to be misinterpreted. In it an attempt has been made to identify a relationship between filter efficiency and air change rate relative to each cleanliness level. It can only be stressed again that it is a guideline for consideration along with the other influences mentioned above, and is not as an absolute panacea.

An important factor in the prevention of particulate build-up within cleanrooms is the use of significant overpressures. In suites of rooms with differing cleanliness levels, pressure gradients can be created, and by subjecting the most sensitive areas to the highest overpressures ensure that the transfer of contamination from room to room is reduced to a minimum. Gradients can be achieved in two ways. Firstly, by careful calculation of air flow volumes each room can be set up to a predetermined overpressure, independently of the others. Secondly, air is transferred from room to room by cascading air through transfer grilles and pressure relief dampers.

The first method reduces the risk of cross-contamination to a minimum but is difficult to balance initially and to maintain in balance if HEPA filters tend to clog at differing rates. Because the second method is interactive it becomes self-regulating once achieved. The system characteristics are readily maintained by increasing the supply air flow rate as filters clog either through a constant volume damper or by a manually operated damper in the main supply duct.

Negative pressure cascades work in an equal and opposite manner with areas of highest sensitivity protected by the lowest negative pressure. The entrance sequence to a containment suite should be protected by a positive pressure zone to prevent a net inflow of dirty air.

2. Air Distribution

There are two generally recognized methods of air distribution within the cleanroom.

The first, turbulent flow (Figure 7.2) is the conventional design approach where terminal outlets represent only a proportion of the total ceiling area located to suit the individual process requirements or general room distribution. Extracts may be located within the ceiling or at low level within the walls. Air flows, whilst predictable, are impossible to guarantee. When HEPA filters are installed in terminal housings, cleanliness levels as high as Class 1000 can be achieved.

For facilities requiring Class 100 and better conditions (Figure 7.3) a unidirectional downflow (laminar flow) air distribution pattern is essential, particularly when 'in use' testing is required. With a vertical air flow of moderate velocity

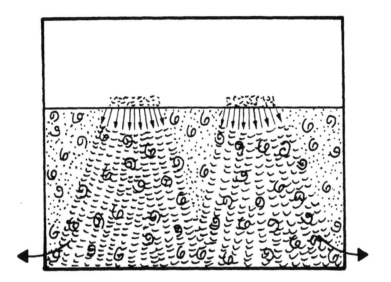

Figure 7.2 Turbulent flow air distribution

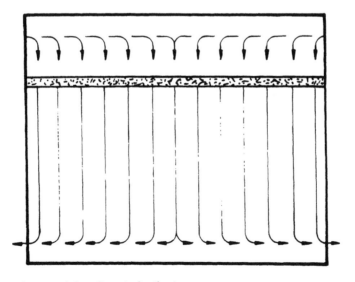

Figure 7.3 Unidirectional downflow air distribution

90 ft/min (0.45 m/s) from a fully filtered ceiling, particle travel is easier to predict, there being no dead areas for contamination build-up. With air change rates as high as 5/600 air changes per hour, both capital equipment and operating costs are significantly higher than is the case with turbulent flow rooms. Wherever practical, laminar flow areas should be restricted, into small rooms, controlled zones or canopies within rooms and self-contained work stations.

As cleanliness levels increase, so the importance of air exhaust locations becomes more significant. Cleanliness levels of Class 100 000 can be maintained efficiently with exhaust air grilles located in the ceiling or at high-level in walls. With higher cleanliness levels, low-level extraction becomes essential. This requires the use of double-skin walls or ductwork located in service chases if unsightly intrusions within the cleanroom are to be avoided.

Predictability of airflow is the ultimate requirement. In this way systems can be designed to protect the product and the operator from the effects of airborne contamination. However, it must be recognized that even in laminar downflow areas the introduction of people, equipment, and the ceiling construction itself into the airstream will result in areas of turbulence which must be minimized (Figure 7.4).

Unidirectional airflow can also be provided, with a horizontal axis. Extensively used in early cleanroom developments, its use is much less popular at the present time, because air quality degrades rapidly the further away from the filter bank one moves. Each operation breaks up the airflow through the room, decreasing predic-

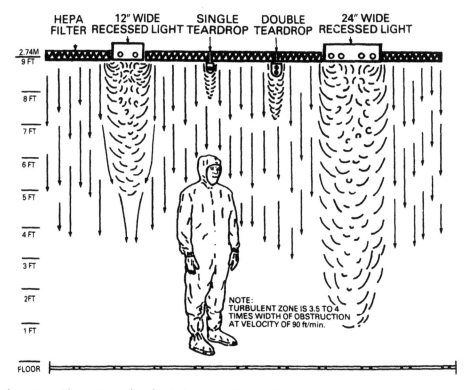

Figure 7.4 Obstruction induced turbulence in a laminar flow cleanroom (Source: American Air Filters)

tability of direction and introducing contamination which is passed on to successive operations.

Horizontal laminar flow is still used extensively in cabinets where product protection is the only consideration.

3. Filtration of Supply Air

The importance of filtration efficiency in achieving cleanliness levels has already been identified (Table 7.2). Cleanroom technology is dependent on the use of high-efficiency particle arrestor (HEPA) filters comprising pleated packs of high-density glass fibre paper with aluminium or craft paper separators, sealed into a timber or metal frame with urethane. Traditionally, filters have been six or twelve inches deep.

Improved manufacturing techniques have resulted in the availability of a range of minipleat filters with a nominal depth of two inches. These may contain up to twice as much paper as conventional six inch filters, achieved by replacing separators with more densely pleated, self-supporting media. The principal advantages are:

- Lower pressure drop, producing reduced system resistance.
- Higher CFM capacity.
- Greater loading capacity, resulting in longer service life.
- Reduced risk of chafing, pinhole leaks, and off-gassing, associated with separators.

Most manufacturers now offer minipleats with various degrees of sophistication.

The relative efficiency of HEPA filters is extremely important, and all filters will be tested at least once during the manufacturing/installation process. There are a number of international standards which set out efficiency and leak testing procedures for filters. IES Recommended Practice IES-RP-CC-0011-83T *HEPA Filters* (1983), BS 3928:1969 *Method of Sodium Flame Test for Air Filters* (1969), and DIN 24184 *Type Test of High-Efficiency Submicron Air Filters*, all specify factory based tests for establishing the integrity and efficiency of filters. Project specifications may require 100% testing or permit random batch sampling. Efficiency tests tend to confirm an overall efficiency but cannot identify potentially harmful pinholes. Whilst providing an overall indication that the specified efficiency has been achieved, they can give no protection against damage during delivery or installation. For this reason specifiers will invariably require an installation leak test to be carried out as part of the system validation. Based on a challenge of injected oil smoke generated from DiOctyl Phalate (DOP) or oil with similar particle size characteristics, such tests not only test the filter integrity but also that of the seals to the filter housing. Requirements for such tests are considered further at the end of this chapter.

Manufacturing improvements have not been restricted to the filter medium. Whilst it has been standard practice to locate filters into permanent housings installed as part of the room fabric/ductwork system using a proprietary gasket or 'fluid' seal, factory sealed cassette filters comprising filter and housing have been widely manufactured in recent years. The principal advantages of factory control over the difficult filter/housing seal and lower initial cost are, however, countered by

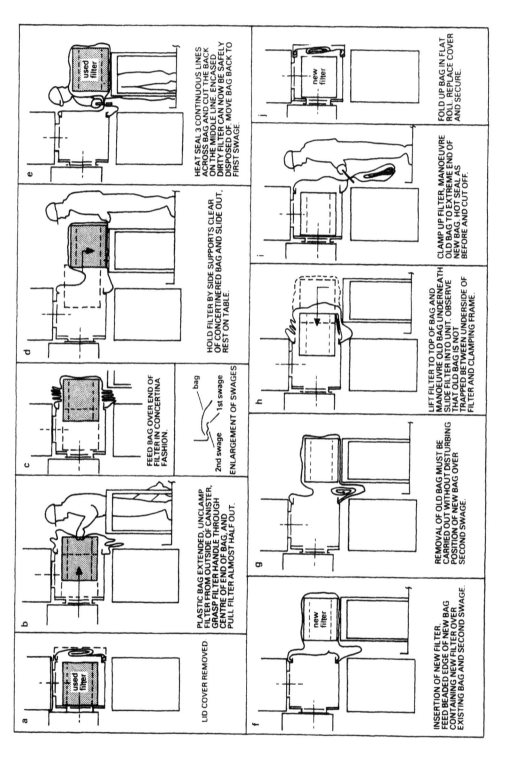

Figure 7.5 Typical safe change filter procedure

a higher replacement cost and a breaching of room integrity during filter changing. The use of cassette filters is not widespread in pharmaceutical areas.

Requirements for containment of potential airborne contaminants within cleanrooms, particularly in biotechnology facilities, may require HEPA filtration on exhaust air systems also. To be effective, filters should be located as close as practicable to the point of extract from the room, reducing ductwork runs susceptible to contamination to a minimum. Filters must be capable of being changed without breaching the integrity of the ductwork system. Known as 'safe change', the filters require special housings and bagging techniques to avoid dislodging and spreading any contaminants collected on the media. A typical filter change technique is illustrated in Figure 7.5.

4. Detail Design and Material Selection

The success of any pharmaceutical construction project may ultimately depend not so much on the process and services provisions which can be evaluated objectively against defined performance specifications but on the perceived 'quality' of the facility based on a subjective evaluation of the finishes. For this reason attention to the selection of materials and the detail of their assembly or application must assume a high priority for the designer.

It is also important in considering structural solutions for a project that in addition to the obvious requirement for suitability, allowance is made for the significant services loads often imposed in a pharmaceutical application. The structure should permit frequent penetration for holes and chases. Care should be taken in design of floor slabs particularly in the location of movement joints to ensure they do not compromise the integrity of applied floor finishes.

Construction Techniques

Choice of construction techniques is influenced by a number of external factors, in addition to the obvious requirements that they be fit for the purpose chosen and do not, of themselves, contribute to the particulate burden. The key factors are:

- Facility location
- Flexibility
- Cost effective solutions.

Facility location: The location of a facility can influence construction materials, not only in terms of availability of raw materials and installation skills, but also by strict interpretation of building codes.

Flexibility: Once particulate contamination performance criteria have been satisfied, and these must not be compromised, flexibility is the greatest single requirement facing facility managers. How to design a facility which meets current performance requirements, can be adapted to accommodate frequent equipment changes and can be upgraded to meet developing standards without requiring total refurbishment.

Cost effective solutions: With so many diverse requirements to be satisfied, a thorough value engineering exercise is required at the detail design stage to ensure that the proposed solutions represent best value per currency unit.

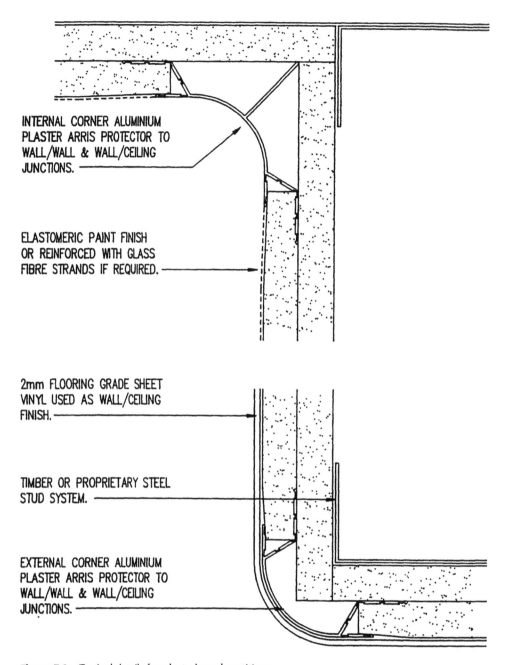

INTERNAL CORNER ALUMINIUM
PLASTER ARRIS PROTECTOR TO
WALL/WALL & WALL/CEILING
JUNCTIONS.

ELASTOMERIC PAINT FINISH
OR REINFORCED WITH GLASS
FIBRE STRANDS IF REQUIRED.

2mm FLOORING GRADE SHEET
VINYL USED AS WALL/CEILING
FINISH.

TIMBER OR PROPRIETARY STEEL
STUD SYSTEM.

EXTERNAL CORNER ALUMINIUM
PLASTER ARRIS PROTECTOR TO
WALL/WALL & WALL/CEILING
JUNCTIONS.

Figure 7.6   Typical details for plasterboard partitions

Primary Construction Materials

With such a wide variety of construction materials and systems available, it has already been suggested that selection can be a matter of personal preference and available budget. Clearly some guidelines can be established to assist selection. The philosophy must be to provide a room fabric which has the following characteristics:

- Hard, impervious, smooth surfaces with no sharp angles or edges, in order to prevent particle generation.
- Coved corners, angles and plinths to facilitate thorough cleaning.
- Smooth, flush details to abutments with openings or the interface of different materials so as not to inhibit air circulation.

Generally, design guides refer to these elements as 'crack and crevice free construction'. Selection of materials will also be influenced by local building codes and in many cases by corporate standards of construction. In the former case, requirements for structural stability or fire resistance and the ability of materials to resist surface spread of flame will be of particular importance.

Corporate standards reflect many years of experience in detailing construction to meet the manufacturing demands of particular product requirement and, invariably, provide cost effective solutions. They form an invaluable starting point for any design exercise.

Externally elevational treatment will be subject to local planning consent and will either reinforce corporate identity or enclose space as economically as possible depending on client policy. We are not so much concerned here with these subjective values but with looking objectively at production spaces within these buildings.

There are essentially three methods of internal wall construction:

- Masonry construction
- Lightweight construction (Figure 7.6)
- Demountable partitions (Figure 7.7).

The two former methods require applied finishes if they are to meet performance standards, while the latter will generally be prefinished. Most in situ construction techniques require the application of a finishing material if the construction is to satisfy design guide criteria. As room cleanliness levels increase or where process demands dictate, so the sophistication of the finishing coat will also increase.

At the lowest end of the scale, this may be as straightforward as water-based vinyl emulsion type paints, applied directly onto fairfaced masonry or gypsum board. The comparative frequency of maintenance can be readily offset by low cost and ease of application.

A considerable range of hard wearing epoxy and similar resin-based paints and spray coatings are available, suitable for the most demanding aseptic applications. Many such products possess an inherent elasticity which resists cracking, even when movement of the substrate induces tension.

Flexible sheet vinyl materials, either 1 mm or 2 mm flooring grade, are often used in aseptic rooms of the highest quality. The vinyl is carried over preformed cove formers at wall/wall and wall/ceiling junctions to ensure easy cleaning. The use of high quality impact adhesives ensures that, even in negative pressure environments, the material stays firmly fixed to the wall. All joints are thermally welded ensuring a smooth, impervious wall surface.

Personal preference may permit the use of other materials. Rigid sheet materials of acrylic and similar substances can, with careful detailing of joints, also provide attractive and hard wearing surfaces. Ceramic tiles, both glazed and unglazed, have

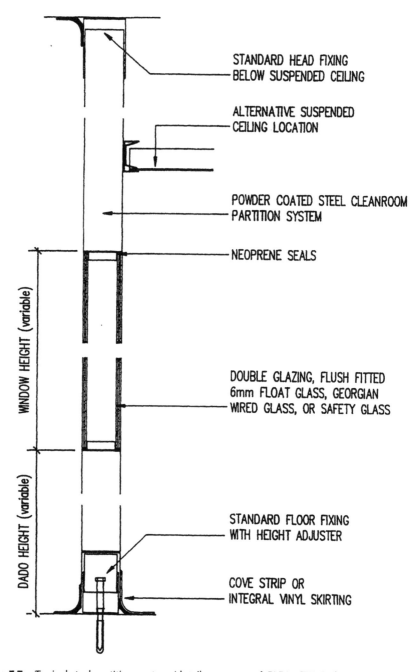

STANDARD HEAD FIXING BELOW SUSPENDED CEILING

ALTERNATIVE SUSPENDED CEILING LOCATION

POWDER COATED STEEL CLEANROOM PARTITION SYSTEM

NEOPRENE SEALS

DOUBLE GLAZING, FLUSH FITTED 6mm FLOAT GLASS, GEORGIAN WIRED GLASS, OR SAFETY GLASS

STANDARD FLOOR FIXING WITH HEIGHT ADJUSTER

COVE STRIP OR INTEGRAL VINYL SKIRTING

WINDOW HEIGHT (variable)

DADO HEIGHT (variable)

Figure 7.7   Typical steel partition system (details courtesy of CLEANTEK GmbH)

been used extensively in the past although their use has declined with the advent of epoxy materials with similar wearing characteristics. The breakdown of grouted joints in more aggressive cleaning environments has made this type of finish extremely suspect.

As with walls, ceiling construction falls into quite distinct categories:

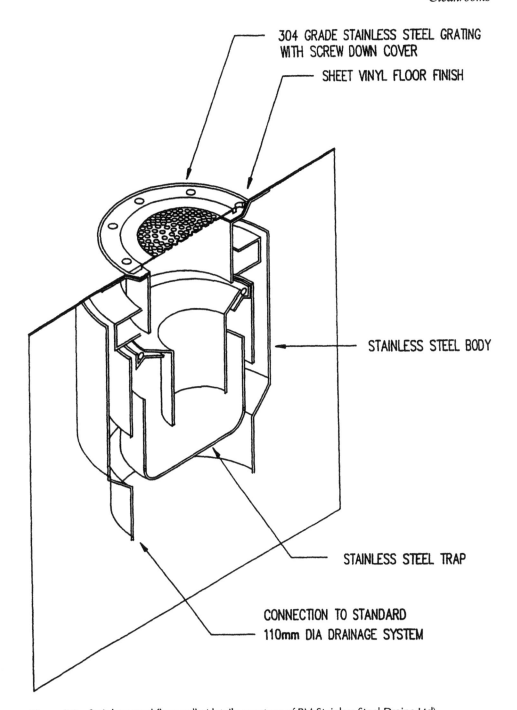

Figure 7.8 Stainless steel floor gully (details courtesy of BM Stainless Steel Drains Ltd)

- Monolithic systems
- Tile and panel systems.

Because of the desire to distribute ductwork, piped and electrical services outside of

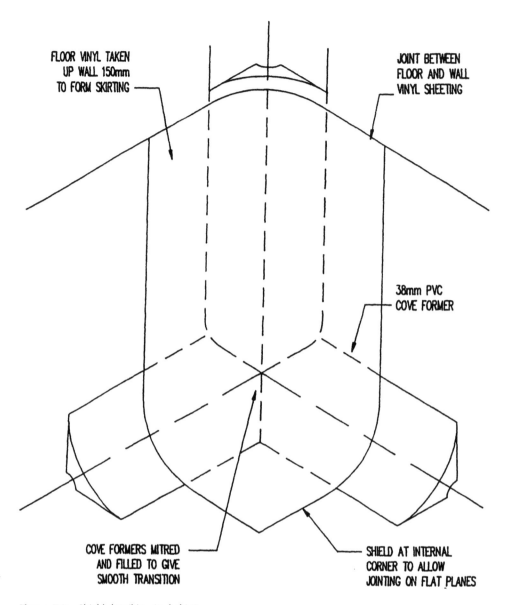

FLOOR VINYL TAKEN
UP WALL 150mm
TO FORM SKIRTING

JOINT BETWEEN
FLOOR AND WALL
VINYL SHEETING

38mm PVC
COVE FORMER

COVE FORMERS MITRED
AND FILLED TO GIVE
SMOOTH TRANSITION

SHIELD AT INTERNAL
CORNER TO ALLOW
JOINTING ON FLAT PLANES

Figure 7.9 Shield detail in vinyl skirting

the cleanrooms, the space above the ceiling becomes an important service zone. Only rarely therefore, will the concrete soffit of a roof or intermediate floor slab be finished as a ceiling. It is much more likely that a ceiling system will be suspended well below such elements, to facilitate service distribution.

Selection of floor finishes will be influenced by local conditions more than any other materials. Of particular importance is the presence of wet processes. Standing water on floors can be a significant problem. Wear characteristics may also be significant if other than pedestrian use is anticipated. Wheeled vehicles, particularly pallet trucks and similar wheeled units, can have a devastating effect on floor materials. In addition to care with selection, detailing is also important with expansion joints, drain outlets (Figure 7.8), wall/floor coved junctions (Figures 7.9 and 7.10) and other material interfaces requiring particular attention.

Materials fall into distinct groupings:

- Sheet and flexible tile materials
- In situ and tile terrazzo
- Floor coating systems: epoxies, paints and seals.

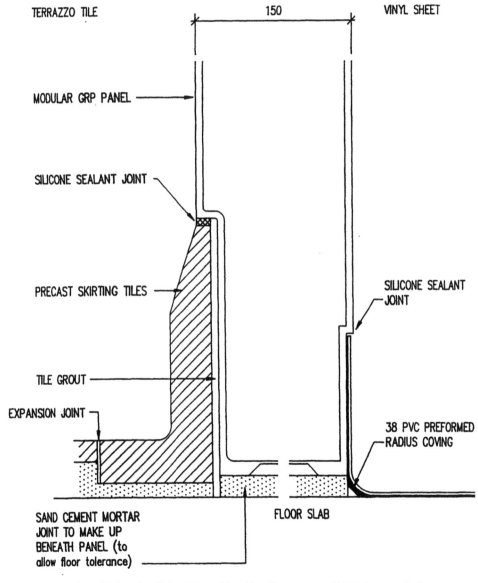

Figure 7.10   Floor skirting details in GRP partition (details courtesy of MRC Systems Ltd)

Secondary Elements

Doors serve two principal functions in cleanroom facilities: firstly to permit the passage of people and secondly to permit movement of materials, either in small quantities by hand and trolley, or in bulk by pallet truck etc. The two functions place quite different requirements on door systems. As cleanliness levels increase, the need to restrict movement in order to reduce the contamination burden assumes increasing significance.

Personnel doors will range from standard painted timber or steel doors in low grade rooms, through solid timber doors which have plastic laminate faces and are edged with hardwood, metal or plastic laminate (Figure 7.11). Doors sets purpose made for higher grade pharmaceutical applications are available in GRP, stainless steel and glass (Figure 7.12). The essential in selecting doors, as with other finishing materials, is the maintenance of hard wearing, crack free surfaces. Attention must be paid to detail design in locating frames into structural openings. Selection of iron-

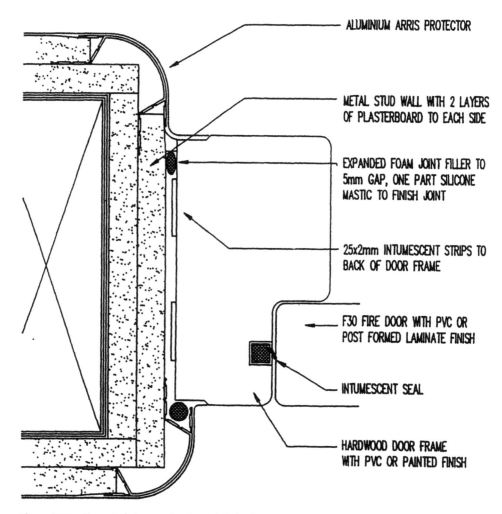

ALUMINIUM ARRIS PROTECTOR

METAL STUD WALL WITH 2 LAYERS OF PLASTERBOARD TO EACH SIDE

EXPANDED FOAM JOINT FILLER TO 5mm GAP, ONE PART SILICONE MASTIC TO FINISH JOINT

25x2mm INTUMESCENT STRIPS TO BACK OF DOOR FRAME

F30 FIRE DOOR WITH PVC OR POST FORMED LAMINATE FINISH

INTUMESCENT SEAL

HARDWOOD DOOR FRAME WITH PVC OR PAINTED FINISH

Figure 7.11 Fire rated doorset (laminated timber)

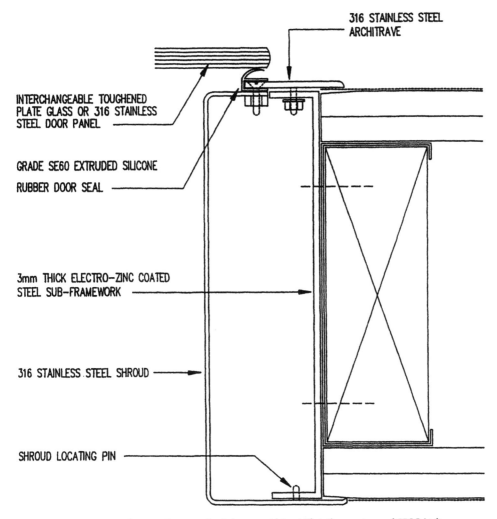

316 STAINLESS STEEL
ARCHITRAVE

INTERCHANGEABLE TOUGHENED
PLATE GLASS OR 316 STAINLESS
STEEL DOOR PANEL

GRADE SE60 EXTRUDED SILICONE
RUBBER DOOR SEAL

3mm THICK ELECTRO-ZINC COATED
STEEL SUB-FRAMEWORK

316 STAINLESS STEEL SHROUD

SHROUD LOCATING PIN

Figure 7.12   Door set for aseptic areas (stainless steel/glass) (details courtesy of PBSC Ltd)

mongery is also important. Closers must work smoothly, often against considerable room overpressure.

It is essential to avoid the use of unnecessary locks and latches in favour of pull handles and push plates. Electromagnetic interlocks, which minimize the penetrations of the door skin, should be used where possible. Additional protection may be necessary on door faces susceptible to damage from truck and trolley movement.

Perhaps because early cleanroom legislation discouraged the use of windows or because satisfactory detailing was difficult to achieve, cleanrooms were, for many years, claustrophobic areas with no natural daylight. There are, however, significant operational benefits from the extensive use of glazing. These might include:

- Greater unity between different sections of the manufacturing process.

- Supervision without the necessity for supervising staff to be continuously entering and leaving the cleanroom through a complex changing process.

- Improved working environment for production operators.

● Greater aesthetic interest.

Glass is, in fact, an extremely suitable material for cleanroom use, as it readily satisfies the principal design criteria, being hard, smooth, impervious and easily cleanable. It can be used effectively in conventional pane sizes (Figure 7.13) or by using thicker, laminated panels for entire full height partitions.

It is therefore important that every effort is made to overcome the technical difficulties, such as contaminated air ingress around frames and solar heat gain. Whenever possible, glazed areas should be flush with adjoining wall surfaces and double glazed, to meet this criteria, on both sides of the wall where the cleanrooms adjoin one another. Glass should be located into purpose made frames in stainless steel or similar materials or located directly into appertures formed in the wall construction using silicone mastic adhesive/sealants.

Material transfers, speech membranes and even conveyor pass throughs can be located with great ease.

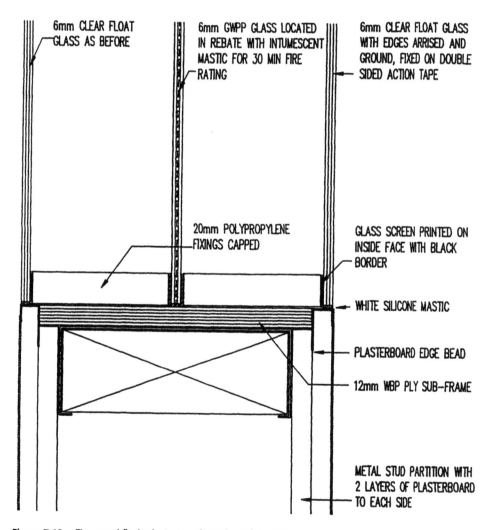

Figure 7.13   Fire rated flush glazing in plasterboard partition

A wide range of fixtures and fittings are essential within the room, if the manufacturing function is to be effective. It is not possible to cover every eventuality, but the list is certain to include:

- Light fixtures
- Filter housings and return air grilles (Figure 7.14)
- Material transfer hatches
- Piped and electrical services (Figure 7.15)
- Autoclaves, production equipment, etc.

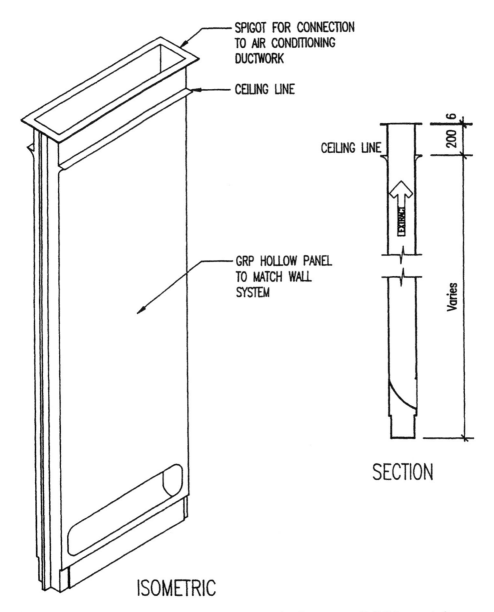

SPIGOT FOR CONNECTION TO AIR CONDITIONING DUCTWORK

CEILING LINE

CEILING LINE

200 6

EXTRACT

GRP HOLLOW PANEL TO MATCH WALL SYSTEM

Varies

SECTION

ISOMETRIC

Figure 7.14 GRP panel with integral return air cavity (details courtesy of MRC Systems Ltd)

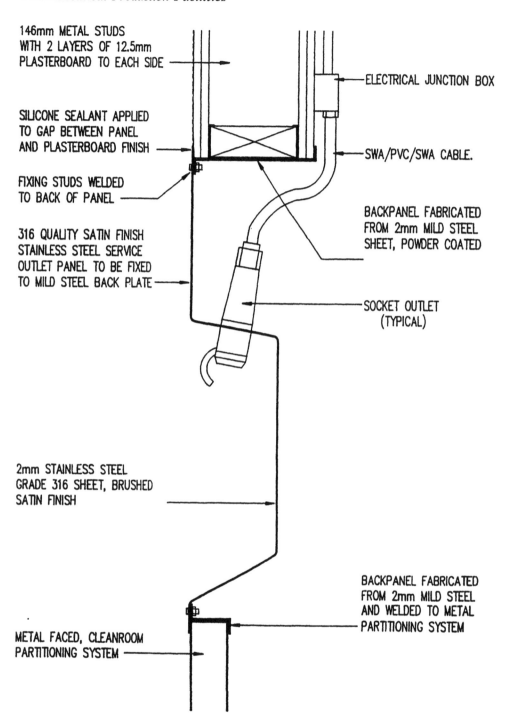

146mm METAL STUDS
WITH 2 LAYERS OF 12.5mm
PLASTERBOARD TO EACH SIDE

SILICONE SEALANT APPLIED
TO GAP BETWEEN PANEL
AND PLASTERBOARD FINISH

FIXING STUDS WELDED
TO BACK OF PANEL

316 QUALITY SATIN FINISH
STAINLESS STEEL SERVICE
OUTLET PANEL TO BE FIXED
TO MILD STEEL BACK PLATE

2mm STAINLESS STEEL
GRADE 316 SHEET, BRUSHED
SATIN FINISH

METAL FACED, CLEANROOM
PARTITIONING SYSTEM

ELECTRICAL JUNCTION BOX

SWA/PVC/SWA CABLE.

BACKPANEL FABRICATED
FROM 2mm MILD STEEL
SHEET, POWDER COATED

SOCKET OUTLET
(TYPICAL)

BACKPANEL FABRICATED
FROM 2mm MILD STEEL
AND WELDED TO METAL
PARTITIONING SYSTEM

Figure 7.15   Utility outlet box

The same considerations given to other aspects of cleanroom detailing will also apply to these components. Wherever practical, fixtures must be flush mounted and sealed into the room fabric. Non-essential equipment should be located outside the room, allowing routine maintenance to be effective without any requirement for maintenance staff to enter the cleanroom, or for the integrity of the room to be breached. Fluorescent light tubes and even HEPA filters can be changed in this way, should service access above the rooms be practical. Where this access is not possible and tubes or filters are changed from inside the room, fixtures must be detailed to ensure that removing diffusers does not breach the room integrity, permitting ingress of contamination.

Long horizontal service runs should also be avoided whenever practical. The zone above the suspended ceiling provides an ideal area for installing service ring mains from where individual services can drop directly to points of use. Services can be grouped and brought into the cleanroom through service pendants. Hollow wall construction may also permit flush mounting of services which would otherwise be an untidy intrusion of the room.

Detailing may be as specific to applications as is function. It is therefore very difficult to provide details which can be regarded as generic or typical.

A significant range of materials and details have been considered here. It is important in concluding to restate the point that production areas must be 'fit for purpose'. It is not suggested for one moment that all processes need to be conducted in rooms that are also suitable for aseptic manipulation. The selection of appropriate materials and construction techniques for differing applications requires the designer to establish a balance between cost and risk: risk that materials may break down and contaminate the manufacturing process and control of project cost within a realistic budget. Establishing this balance and applying similar consideration to development of layouts based on a thorough understanding of manufacturing process and utility requirements are essential contributions to be made by the design team towards the creation of a GMP manufacturing facility.

5. Operational Procedures

The final aspect of the definition quoted earlier in this chapter concerns operational procedures within a cleanroom facility. It is particularly important as an aspect of the overall problem of contamination control, because it is beyond the influence of the designer/contractor and entirely the responsibility of the user.

Having established that point, the following paragraphs examine several aspects of cleanroom operation as a contribution to discussion of the overall problems.

Cleanroom Clothing

It is often stated that the single biggest source of contamination in the cleanroom, is the people who work there. However, this problem is assuming an ever decreasing importance because there are now available to the user a range of garments which offer substantial protection to the product.

Table 7.3 IES-RP-CC-003-87T cleanroom clothing applications

Cleanroom class	Coveralls	Headgear	Footwear	Coats/ frocks	Hands
1	Yes	Full hood and mask	Long overboots	No	Powder and lint free
10	Yes	Full hood and mask	Long overboots	No	Powder and lint free
100	Yes	Full hood, mask as required	Long overboots	No	Powder and lint free
1000	Yes or coat	Hood or snood	Overboots or overshoes	Yes or coverall	Powder and lint free
10 000	Yes or coat	Hat or cap	Overshoes	Yes or coverall	As required
100 000	Not required	Hat or cap	Overshoes	Yes	As required

It may be noted that the above recommendations are based upon a density of one person per 100 sq. ft. of floor space

The correct use of cleanroom clothing by the pharmaceutical industry is becoming more relevant and receiving greater emphasis than before, as cleanrooms themselves operate at high levels of performance for both product integrity and operator safety.

It goes without saying that cleanroom clothing must protect the environment from the wearer and should be designed and produced to meet the highest standards.

State-of-the-art facilities require the use of one-piece, coverall suits, normally with integral hoods, knee-length overboots, and gloves.

It is not uncommon for cleanroom operators to supply two-piece antistatic poly-cotton undergarments and slippers for use generally by staff within the facility, allowing street clothes and outdoor footwear to be stored in change rooms close to the building entrance. Face coverings are also important.

Fabrics must be made from 100% synthetic continuous filament polyester and be constructed to act as filters as well as to be inherently low linting.

If inherently antistatic fabrics are used, care should be taken to evaluate the actual antistatic performance of the garment as well as its basic ability to contain particulates. Antistatic grids can be incorporated in the weave if necessary.

Table 7.4 IES-RP-CC-03-87T cleanroom clothing recommended usage

Cleanroom classification					
1* Each entry	10* Each entry	100 Daily†	1000 Daily	10 000 Daily	100 000 Each other day

† Sterile or aseptic suites conform generally to Class 100, US 209e and Class 1E.BS 5295. Garments are normally changed on each entry.

Garments should be designed and ergonomically engineered to fit correctly with openings sealed to reduce emission, and the whole garment being loose enough to reduce internal abrasion and pressure build-up.

Wearer comfort is the fourth element in deciding on the optimum design/fabric/garment mix.

The Institute of Environmental Sciences, as part of the revision of Federal Standard 209, has produced a Recommended Practice, RP-CC-003-87T *Garments required in Clean Rooms and Controlled Environment Areas* (1987), setting out the specification requirements for garments and their scope in use. It also deals with the frequency with which they should be changed and requirements for laundering.

Tables 7.3 and 7.4 summarize the key points of this document.

Sanitization

The manufacture of sterile or aseptic pharmaceuticals will require the frequent monitoring of production areas in order to guarantee that their contribution to the contamination burden within the process is kept within acceptable limits. Regular and careful janitorial activity using cleaning formulations such as sodium hypochlorite is an important part of this process and can be tested on a regular basis by QC analysing swabs or settle plate samples taken within the room. However, whether as a matter of routine or as a safety procedure after leaks or spillage of active products, it may be necessary to take more stringent measures. The most widely used method is gassing with formaldehyde. This substance is lethal not only to bacteria but also to human beings, and the safe use and disposal of these substances must be considered early in the design process.

The use of formaldehyde for 'dynamic' gassing of rooms has been tried but largely discontinued. This required gas to be injected into rooms either directly or via air ductwork systems. The gas is, however, inherently unstable and requires very tight temperature tolerances throughout the air conditioning (AC) system which circulates the gas. The integrity of ductwork and AC plant which must circulate the gas under significant positive pressures is also problematic and beyond normal manufacturing tolerances for such equipment. Throughout the sterilization period, doors, conveyor apertures, and similar room penetrations must be sealed to prevent leakage. This is best carried out by using a highly visible method such as adhesive safety tape to avoid any possibility of personnel straying into areas during the gassing cycle. When dynamic gassing is to be considered the designer and operator must be aware of the dangers of any leakage from the system into other parts of the facility. Gassing should be undertaken only when the facility is empty, at weekends for instance. Wherever practicable, separate air conditioning systems should be provided to each sterilizable suite of rooms to minimize the risk of accident.

Passive gassing is by far the most widely used method of formaldehyde sterilization. For this method a solution is evaporated within a room where the air systems have been turned off. The gas circulates within the room by natural convection, being allowed to contact all room surfaces. After a designated period the rooms must then be purged and the air handling equipment which may normally recirculate a high percentage of air must have a capability of supplying 100% fresh air and dumping 100% exhaust air during this period. System control is extremely important at this time to avoid overpressurizing rooms and permitting escape of gas through room fabric.

Validation

Before production can commence in a newly completed cleanroom facility, it must be licensed along with all other aspects of the manufacturing process by the regulating authority within whose area the pharmaceuticals are intended for sale. The licence will be granted in respect of the cleanroom only after presentation of validation data confirming that the facility performs in accordance with the design brief in every respect. The subject of validation in the context of the total process is covered in detail in Chapter 15. It is, however, appropriate that the concluding remarks of the present chapter should also touch briefly on the subject. This is not to imply that validation requirements need not be considered before the conclusion of the project, though regrettably this is too often the case.

Whilst cleanroom designers and constructors are used to commissioning and setting their projects to work there is only limited understanding of the legal obligations inherent in the obtaining of a pharmaceutical manufacturer's licence and the degree to which all aspects of the process and its support activities must be documented.

As soon as the process requirements are defined it is essential that a Master Plan is prepared. All elements of the process equipment, utilities, or facilities which have product contact are involved in the validation exercise. Many companies go far beyond statutory limits to satisfy their own good housekeeping standards. The design brief will have identified the cleanliness level at which the room is required to perform. Successive stages of the design process will clarify further detail, and each step will answer questions with increasing confidence that the completed project will satisfy requirements.

Design Qualification

If the performance of cleanroom systems is to be properly tested, this must be against the criteria established as part of the design process. It is important therefore that the design documentation is presented in a co-ordinated manner and can be checked against the user's initial requirements for the project. The Front End Engineering Study document is frequently used as the design qualification (DQ) and would normally produce the document in such a way that it includes a signature page so that it can be taken on board by the Validation Manager, QA, Production etc. as part of the formal documentation of the project.

In addition at this stage, the format of the overall validation documentation will be established. It is important that the design team is aware of the format to ensure compatibility during the subsequent detail design work.

Validation affects two essential areas of cleanroom construction, the layout and the finishes.

In general, it is vital that the rooms are laid out in the correct relationship with each other. They must comply with the process flow for each process, segregation requirements to prevent cross-contamination, personnel flow, personnel segregation requirements to prevent cross-contamination, and access for maintenance in a separate zone to the process.

Problems can occur with these elements when there is no clear validation under-standing, defining, for example, the need to segregate different standards of oper-ative (e.g. black or grey).

Validation requirements also affect the specification of the air quality standards, air distribution patterns, cascade pressurization requirements, filtration require-ments, etc., which must all be defined at this stage.

This document is chosen because it is likely to include a clear statement of the user's requirements usually referred to as the scope definition. This will either be in the form of a design brief offered by the user to the design team or will have been developed as a result of initial research carried out by the design team through examination of existing facilities and by interviews carried out with user depart-ments. Typically the document will contain the following information.

- Room data sheets
- Room layouts
- Process flow sheet
- Equipment lists
- Cleanliness regimes and other environmental considerations
- Utilities and special services provisions
- Operator safety and fire precautions evaluation
- Flexibility analysis
- Adjacency matrix.

It will also include what is likely to be the first clear statement of the design intent. Where projects are implemented on a fast track basis, and increasingly this is the case, it may be the only point where a comprehensive design package is available at a consistent level of detail.

The documentation can be summarized as follows:

- Architectural general arrangement drawings (preliminary plans, elevations and sections at 1.100 scale)
- Preliminary building frame, foundation and other structural details
- HVAC schematics and environmental definition (cleanroom classifications etc.)
- Electrical schematics
- Utilities and special services schematics
- Preliminary finishes schedule
- $\pm 10\%$ cost estimate
- Outline construction programme.

Now more than ever before, a philosophy of validateable design prevails, requiring of the designer considerable experience in pharmaceutical practice and a total understanding of the validation process.

Successive stages of the design process will clarify further detail; each step will answer questions, increasing confidence that the completed project will satisfy requirements.

- What spaces are served by which HVAC systems?

- What is the philosophy towards recirculated air?
- Are there requirements for control of dust or other hazardous materials?
- Is cross-contamination a potential problem, and if so, how should systems be segregated?
- What are the requirements for control systems?
- How will routine monitoring be performed?
- Can a planned maintenance programme be implemented without breaching the integrity of manufacturing spaces?
- What limitations will the process impose on selection of construction materials?

These are questions routinely considered by designers on any project. In this context, each must be considered for the impact it might have on facility validation.

At the detail design stage, design of the specification must be more thorough for pharmaceutical facilities than perhaps is considered normal in the construction industry. This raises the question of how specifications themselves are validated.

Despite such care, it is still possible to quote examples where implementation of a particular specification has created validation problems. A popular choice with HVAC designers in recent years has been the cassette type HEPA filter, where filter media and housing are a single factory assembled and tested unit. An excellent choice in theory! The factory test of sealed housing and media overcomes the most troublesome interface for site installations between filter housing and compressible gasket. There are significant cost benefits also for initial capital expenditure, if at the expense of a small premium on future replacement costs. Installation, however, can be a nightmare! How can the room be cleaned up to accept filters when the fabric of the room is not complete because housings have not been located? How can air systems be blown through to receive filters when they in turn are not complete? How are the constant volume dampers for instance, to be fully commissioned when the system cannot be pressurized? The end result is considerable additional temporary works and inconvenience, ceiling closures, temporary ductwork terminations, etc., the costs of which may outweigh the original benefits.

We might frequently criticize designers for neglecting buildability but in reality, a majority are totally remote from validateability!

How many commissioning engineers have suffered torture and probably injury because a design did not permit adequate access to tests points? How many air systems have been loaded with vast quantities of DOP for hours on end because the designer never specified insertion points on individual filter housings?

A principle of 'clean construction' must be applied throughout the installation process. Designers must recognize for example, that any requirement for ductwork to be cleaned and sealed before delivery to site and for the system to be maintained in a sealed condition throughout its installation period and beyond is a fundamental part of the specification. Users must recognize that 'clean construction' techniques are not a standard part of the construction process and will invoke cost premiums. The benefit of course, is a smoother progression towards the commissioning and validation exercises.

Physical testing will fall into two categories within the Master Plan and these are now considered.

Installation Qualification

Installation qualification (IQ) is the checking and certification that all aspects of fabric and systems installed in the cleanroom meet the specification standards set out in the approved DQ document.

Where any element of the project has changed from the design intent stated in that document, a Change Control Notice must have been issued and signed by all the signatories of the DQ. It must state:

- What the change entails
- Why the change was necessary
- The anticipated impact of the change on programme and cost.

The requirement is to provide an auditable paper trail from the DQ to the As Built documentation.

The IQ will see the gathering together of much documentation including certification of factory performance tests on equipment such as air handling units, pressure tests on ductwork during installation using recognized industry standards such as DW142/3 (1983), and certificates of conformity for materials, particularly those in contact with the product. An example is the stainless steel in pipework and in equipment parts. It is no longer adequate merely to state that it is fabricated in 304 or 316L grade stainless steel. Certification is required from the manufacturer of the steel quoting production batch numbers etc., which prove by testing that the material is adequate for its intended purposes.

It has always been a requirement that welders of stainless steel pipework for instance must be qualified by a recognized agency for the type of work undertaken and must submit daily test welds for analysis. The weld log for each operator forms an integral part of the documentation. We are increasingly, however, being asked to submit CVs for all construction operatives from labourers upwards, identifying that they have been properly trained within their sector of the industry for the work they are employed to undertake and that they fully understand the significance of that work in the context of the current project.

It goes without saying that the design team will have been similarly prequalified.

Operational Qualification

By the time this aspect of the validation exercise is reached, qualification of much of the fabric of the cleanroom has been completed. Systems such as HVAC, lighting and power installations are also completed and operational. The operational qualification (OQ) document records the tests which demonstrate that these systems meet the performance levels specified in the DQ.

Again, recognized industry standards must be applied, i.e. CIBSE commissioning code A (1971) for air conditioning systems, 16th edition of the IEE regulations for electrical wiring, etc (Institute of Electrical Engineers). Air volume tests, extract volume tests, temperature and humidity tests, illumination levels, noise and vibration tests will all have to be checked. Load tests will have been conducted on fan motors. Ductwork will have been pressure tested prior to lagging. All tests will also have been witnessed and documented.

Specific cleanroom testing will form part of the OQ. The scope of these tests are identified in documents such as BS 5295:1989 Part 1 or IES-RP-CC-006-84T. They will include:

- Differential pressure test
- HEPA filter installation leak test
- Fabric induction leak test
- Particulate contamination test.

Other tests required by the previous standards are no longer mandatory but can, at the user's insistence, still be carried out in laminar flow environments for instance. These include:

- Airflow uniformity test
- Airflow parallelism test.

A key element of the documentation will be the collection of all relevant design and vendor data. This is an area that often causes problems as a result of the reluctance of manufacturers to provide specific information about their equipment at the appropriate time, relying on more general technical literature that may not be completely accurate with regard to the projects specific equipment.

It will be necessary to ensure that complete and accurate technical literature is available for use by the design team during the detail design and construction processes. Again, BS 5295:1989 identifies specific requirements to be included on test certificates and this fits neatly into the overall documentation requirements. These would normally include:

- Applicable standard
- Title of the test quoting protocol
- Classification
- Test equipment used including serial number and copy of calibration certificate
- Test result
- Special conditions occuring during the test
- Signature of test engineer and company
- Acceptance signature
- Back-up data including location plans, equipment printouts, etc.

It should be remembered that the format of the documentation will already have been established by the Master Plan protocols, so that the information can be presented as a co-ordinated whole.

References

IES Recommended Practice IES-RP-CC-006-84T *Testing Clean Rooms* Institute of Environmental Sciences, Mount Prospect, Illinois 60054, May 1984.
Guidelines for Research Involving Recombinant DNA Molecules, Notices Department of Health and Human Services, Federal Register Part III Vol. 51 No. 88, 7 May, 1986.
ACGM/HSE/Note 6 Advisory Committee on Genetic Manipulation *Guidelines for the Large-scale Use of Genetically Manipulated Organisms.*

BS 5295:1989 *Environmental Cleanliness in Enclosed Spaces* Part 0, 1, 2, 3, & 4 British Standards Institution, London, England 1989.

IES Recommended Practice IES-RP-CC-001-83T *HEPA Filters* Institute of Environmental Sciences, Mount Prospect, Illinois 60054, 1983.

BS 3928:1969 *Method of Sodium Flame Test for Air Filters* (*other than for air supply to IC engines and compressors*) British Standards Institution, London, England, 1969.

DIN 24184 *Type Test of High-efficiency Submicron Particulate Air Filters*. Deutsches Institüt für Normung V, Berlin, West Germany.

IES Recommended Practice IES-RP-CC-003-87T *Garments required in Clean Rooms and Controlled Environment Areas* Institute of Environmental Sciences, Mount Prospect, Illinois 60056, USA, 1987.

DW/142 *Specification for sheet metal ductwork low medium and high velocity systems* Heating and Ventilating Contractors Association London 1983.

DW/143 *A practical guide to ductwork leakage testing* Heating and Ventilating Contractors Association, London, 1983.

Commissioning Codes, Series A, *Air distribution systems, high and low velocity* The Chartered Institute of Building Services Engineers, London 1971.

Regulations for electrical installations 16th Edition, Institute of Electrical Engineers, London.

8

Tablet Production Systems

... a shower of little pebbles came rattling in at the window ... .

The development of tabletting equipment during this century has been largely one of continuing evolution, apart perhaps from exceptions such as presses designed to produce coated or layered tablets. In many areas the incentives have come from the pharmaceutical industry (rather than the press manufacturers) as a result of certain trends in tabletting operations. These include a desire for higher rates of production, direct compression of powders, stricter standards for cleanliness as part of an increasing awareness of GMP, and a wish to automate, or at least continuously monitor, the process. More recently the recommendations of the various bodies have provided an additional impetus to innovation. However, there is now evidence to suggest that the inherent limits to further development of some press variables on existing lines are now being approached.

At present, and in the immediate future, there will be the continuing vying of one manufacturer with another over relatively minor improvements, with maybe more significant advances in the field of instrumentation and automatic control. There is of course the added possibility of some major design development, but the indications are that this will be a rare event. This chapter will concentrate on the current evolutionary innovations and those of the recent past which are meeting the challenge of modern tabletting departments. Tablets are probably the most convenient dosage form there is, and there are many different types of tablet, round, shaped coated, and sustained release, to mention but a few.

The basic design of tablet making equipment has not changed for well over a century. Although many alternative methods have been tried, the principle of filling granules into a die and pressing them into a tablet between two punches is still the exclusive method of manufacture for all machines used in the pharmaceutical industry. It is not intended here to provide a treatise of tablet making technology but to illustrate some of the control systems that are in use and their impact on the automated facility concept.

To manufacture tablets it is necessary to consider a whole range of substances other than the drug. The traditional tablet making process consists of mixing the constituents of the tablet, i.e. the drug, the excipients such as binders, filler, colouring material, etc., and forming the mixture into granules. The purpose of granulating is to prevent segregation of the drug and to produce a free-flowing material which will facilitate consistent die filling and hence give consistent tablet weights.

Some formulations may be dry mixed and fed directly to the tablet machines.

125

This process is known as direct compression. Care is needed in formulating direct compression mixtures to avoid segregation which can result in an inconsistent drug content of the tablets. Special binders are frequently required, but their cost can be offset by the savings made by eliminating the granulating process. A number of direct compression mixtures are in current use because of the difficulty of devising suitable formulations.

The granules or direct compression mixture can be compressed on a variety of tablet making equipment. All tabletting presses employ the same principle; they compress the granular or powdered mixture of ingredients in a die between two punches, the die and its associated punches being called a station of tooling. Tablet machines can be divided into two distinct categories on this basis:

- those with a single set of tooling – 'single station' presses or eccentric presses
- those with several stations of tooling – 'multi-station' or 'rotary' presses.

The former are used primarily in an R&D role, whilst the latter, having higher outputs, are used in most production operations. The most suitable machine is selected to suit the type of tablet to be made and the output required.

Additionally, the rotary machines can be classified in several ways, but one of the most important is the type of tooling with which they are to be used. There are two types of 'B' tooling which is suitable for tablets up to 16 mm diameter or 18 mm maximum dimension, depending on shape if the tablet is not round, and 'D' type which is suitable for tablets with a maximum diameter or maximum dimension of 25.4 mm. The 'B' type punches can be used with two types of die; the small 'B' die is suitable for tablets up to 11.1 mm diameter (or maximum dimension) and the larger 'B' die is suitable for all tablet sizes up to the maximum for the 'B' punches. Machines are consequently suitable for either 'B' punches or 'D' punches but not both, and the machines taking 'B' type punches are designed to accommodate either the large or small dies but not both. Machines taking 'B' type tooling are designed to exert a maximum compression force of 6.5 tons, and machines taking 'D' punches 10 tons. Special machines are available which are designed for higher compression forces. The maximum force which can be exerted on a particular size and shape of tablet will be governed by either the size of the punch tip or the maximum force for which the machine is designed, whichever is smaller.

The tablets can either be packed directly after manufacture or they may require coating for a variety of reasons such as protection against atmospheric oxidation, to counter unpleasant taste, or to improve their appearance. In some cases the tablets, when coated, improve the packaging operation, i.e. blister packing, and in others reduce problems of staining patients' clothes. The traditional method of coating was to apply a sugar coat which typically doubled the tablet weight. More recently film coatings of cellulose polymers or similar materials have been used. They are equally effective, and by reducing the weight of the finished tablet, reduce packing and transport costs. A more detailed description of the coating process appears in a later chapter.

Tablet Weight Consistency

Having produced either a direct compression mixture or a granulation, with the drug distributed evenly throughout the product, the next stage is the production of a consistent dose for the patient. It is, therefore, necessary to consider all the factors which affect the weight of the final tablet.

Rotary Tablet Making Machines

A compression cycle is shown in Figure 8.1. In this type of machine the operating cycle and methods of filling, compressing, and ejection are different from those of the single station presses.

Some of the more important parts of the machine are:

1. Pressure rolls – should be accurately concentric, otherwise random unevenness in tablet thickness will be produced. In modern high-speed presses, play between the roll and its pin is undesirable because more than one punch will be in contact with it at any one time. Dwell times are increased by using rolls of larger diameters in some high-speed presses.
2. Weight adjustment cam – the end of the ramp must coincide precisely with the end of the feeder which scrapes away the excess material; worn ramps can be a

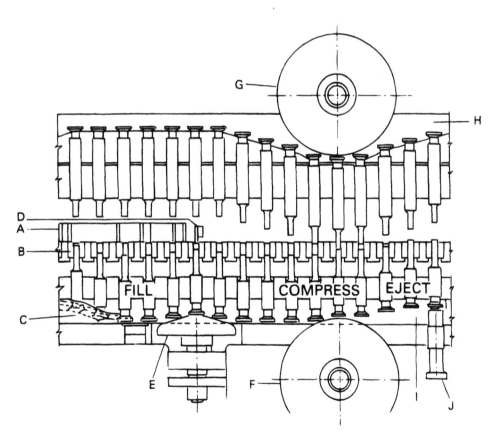

Figure 8.1 Compression cycle of a rotary press. A Feed frame. B Die. C Pull down cam. D Wipe off blades. E Weight control cam. F Lower compression roll. G Upper compression roll. H Upper raising cam-upper punches. I Lower raising cam-lower punches. J Ejector control

source of weight variation, and punch 'flighting' can happen in high-speed presses.

3. Ejection cam – forces can be quite significant here with the punches sliding up it, and for this reason the cam is commonly made of a softer material than the tooling and therefore wears in preference to them. In high-speed machines punch 'flighting' can again occur here, unless it is deliberately prevented by a 'pull-down' cam, which also overcomes any tendency for punch sticking.

4. Feed frame – the traditional method of feeding material from a stationary feed frame falling under the influence of gravity into the die cavities, the feed frame being designed to pass material to and fro across the die surface. More sophisticated types are needed for high-speed machines, and these do not rely on gravity, but are mechanically operated. Close clearance with the die table is essential to avoid generation of excess fines, but must be sufficient (in some presses) to allow release of fines.

The ways in which individual manufacturers of equipment have sought to achieve higher output fall into four groups:

- by increasing the effective number of punches, i.e. multi-tipped types
- by increasing the number of stations
- by increasing the number of points of compression
- by increasing the rate of compression, i.e. turret speed.

Each of these approaches has its own particular advantages and disadvantages, but in addition, all make demands on other aspects of press design, and certain general inherent characteristics of die compaction have had to be taken into account.

Generally the high-speed machines consist of 'double rotary' presses where the cycle of operation is repeated twice in one revolution of the 'turret' carrying the tooling. These machines normally have odd numbers of stations, up to 79 in the largest presses. 'Double-rotary' presses can also be modified to produce layered tablets, whilst other machines have been adapted to produce coated tablets, by a 'dry' compression technique.

The ability to compress a particular product on presses with different output rates, without affecting the time for the actual compaction process, is especially attractive, for at least three reasons. The first is that of the flexibility offered by being able to run a single formulation on different machines. But probably more important is the circumvention of difficulties associated with some minimum compaction and recovery time, related to the deformation characteristics of the major components and the need to allow a finite time for the displacement of the air inevitably present in the feed material. In view of these factors it is perhaps surprising that the use of multi-tip tooling has not been more popular, with the exception of Europe where up to 6 tips per stem are now offered by some manufacturers. The higher initial and maintenance costs of such tooling are the most likely explanation for any reticence, plus the increased restrictions on tablet diameter. For the former reason the introduction of tooling where the individual tips on a particular punch stem are readily replaceable has immediate appeal, although matching and maintaining constant overall length of each punch and tip demands a good tool maintenance facility. This is a comment which has to be more generally applied as tabletting operations become more sophisticated.

Reducing the entire sequence of the compressional cycle into half the turret periphery as a means of virtually doubling output was realized in the first 'double-rotary' presses just after the turn of the century. This concept has now been taken a stage further in a 'four-sided' press in which the entire compression cycle is restricted to one quarter of the cycle and repeated to give a total of four tablets from each station every revolution.

One development which attempted this was the Magna Press, manufactured by Horn, but it never reached commercial success. Developments like this have effectively reduced the time available for die filling, and considerable effort by press manufacturers has been directed toward improved feeders. A further incentive here has been the continuing clamour from the pharmaceutical industry for presses capable of handling a wide range of powdered feed materials.

Press design must always bear in mind the inherent relationship between the centrifugal force 'F', the diameter 'D', and speed 'n' (rpm) of the revolving turret:

$$F = k.D.n. \ldots, \tag{8.1}$$

where F is in multiples of the gravitational force and in high-speed presses may be as much as $10 \times$ that of first generation rotary machines.

Possible movement of material in the die, due to centrifugal effects, can be largely overcome by a precompression stage in the cycle, unless material is actually being lost from the die cavity. Then it may be necessary to take advantage of another innovation, 'variable punch penetration', and carry the compressed material lower in the die or virtually seal off the cavities between filling and compression points as in some Stokes presses. Loss of material was not the main reason for the development of variable punch penetration, which really arose from attempts to improve tooling life by compressing the tablet near the top of the die cavity.

A further problem associated with increased turret speeds is the change of punch velocities along the cams, particularly the weight adjustment cam, exceeding some critical value, leading to a phenomenon known as 'punch flight'. The machine modification of fitting spring-loaded plungers which pressed against the punch body, originally developed as 'anti-turning' devices, is some help in this respect.

The latest presses are fitted with 'generated cams', that is cams whereby all punch accelerations are under control and at acceptable levels.

With one or two notable exceptions the development of tabletting equipment has been essentially a continuing evolutionary process. However, in recent years pressure from the pharmaceutical industry, largely for presses with higher outputs and better performance, has led to an increased impetus in this process.

In addition to the standard rotary tablet machine there are a number of special compressing machines designed to produce dry granulations, ('slugging'), compression coated tablets, layered tablets, and effervescent products.

The Monitoring and Control of Tabletting Processes in Large-scale Production

The manufacture of tablets by high speed machines requires strict control of tablet weight to ensure a uniform dosage form. The rationale of compressing volumetric samples of a granular material fed from a hopper leads to weight variation. To achieve a satisfactory level of tablet weight uniformity three main conditions must be satisfied:

1. The material to be compressed must flow evenly from the hopper into dies to ensure uniform volumetric fills of the die.
2. The packing within the dies must be uniform. As peripheral speed increases so the centrifugal forces acting in the material in the dies result in uneven packing.
3. The dimensions of the machine tooling must be within very tight limits. Variations in granule flow, granule size distribution, or machine performance may lead to drifts in mean tablet weight or excessive weight variation. These can affect other physical parameters of a tablet's characteristics such as disintegration, hardness, and thickness which in turn affect such important parameters as dissolution.

In-line weight control can be divided into three broad categories:

1. Manual sampling and weighing.
2. Closed loop automatic control.
3. Semi-automatic systems based on microprocessors or computer linked systems.

Control of Tablet Weight

The traditional method of controlling tablet weight was by the operator taking samples at fixed time intervals (usually 10 tablets every 15 minutes), weighing to obtain an average weight per tablet: then comparing this figure with the required tablet weight and adjusting the machine as necessary. A further check was then made by the quality control department on samples taken from the batch. In these tests individual tablets are weighed to ensure that the batch meets the required standard. The various pharmacopoeia recommend the weighing of 20 individual tablets, 18 of which must be within certain limits, and the other two within slightly wider limits. The limits are usually given in percentages of the mean weight and vary with the mean weight of the tablets, larger percentage deviations being allowed for smaller tablets. It is frequently found that pharmaceutical manufacturing companies usually set in-house limits much tighter (typically half) than those set by the various pharmacopoeia. This is to ensure that if the sampling technique misses tablets which have a slightly wider deviation than normal, the whole batch has an exceptionally good chance of being within the pharmacopoeia standards.

Increasing machine speeds have created a need for more frequent checking of tablet weight to ensure that the batch complies with the required specification. This, coupled with the desire to reduce the labour content, both from a reliability and cost point of view, has led to the need for automatic weight checking. An electronic system which emulates the operator in taking a sample at a predetermined time interval, weighting the individual tablets and then calculating the mean weight, was introduced approximately 15 years ago by C.I. Electronics. This system could then print and/or store the results and either provide an indication to the operators for them to make the adjustment to the machine if necessary, or it could be arranged to adjust the machine automatically. This system is still available, but does not appear to have found much favour in the industry.

The alternative method is an indirect means of monitoring tablet weight. Most tablet making machines are designed to produce tablets of constant thickness; consequently small changes in the amount of material used to form the tablet (i.e. variations in weight) give rise to corresponding variations in the force used to compress the tablet. These forces can be monitored by means of strain gauges or piezoelectric force tranducers, the mean force/weight established, and variations from the required mean used to adjust the mean tablet weight. One variation to this system exists where a particular machine manufactured by Courtoy makes tablets at a constant force. The variations in thickness, which result from variations in the weight of material used to form the tablet, are then used to maintain the required mean weight in a similar way to that used by the force measuring system.

Both these methods can only be used to control mean tablet weights. The force or thickness can be measured only when the tablet is made, and at that stage it is too late to alter the weight of that particular tablet. Other factors which can interfere with this method of control on multi-station presses (which are essential for high-speed production) are:

A. variations in punch length
B. variations in punch tip geometry
C. variations in die bore
D. variations in freedom of punches to move within their guides (tight punches)
E. variations in the bulk density of the feed material.

A, B, and C can be reduced to a minimum by careful selection of the punches and dies within a set and regular checking to ensure that uneven wear does not occur within the set. D can be overcome by electronic detection if the force deviates sufficiently to affect the accuracy of the control system. However, at the present time there is no equipment available to continuously monitor E, the bulk density of the feed material.

In spite of these problems most tablet machine manufacturers can provide force (or thickness) based control systems, and as variations in the bulk density of the feed material to tablet presses do not occur very often they are, in the main, successful in monitoring or controlling tablet weights at high speed. These units have been so successful that they are frequently fitted with printers (e.g. Manesty Micro P) which produce records used as batch documentation. These records are often considered as proof of weight consistency. However, the force/weight relationship cannot be guaranteed, and this is a dangerous practice.

One of the main advantages of these systems is that they are able to check every tablet, and while the individual force measurement may not relate accurately to the tablet weight, very high or very low force readings from an individual tablet can at least indicate that the tablet is suspect. Some tablet machine manufacturers offer equipment to reject these individual tablets, which can either be discarded or reworked, if of low value, or individually sorted on an automatic balance, if their value warrants this action.

The systems of sample weighing and force or thickness measurement have both advantages and disadvantages. The weighing system has the advantage of measuring the actual parameter which it is required to control, but does not have the ability to check every tablet. Conversely, the force or thickness measuring system is capable of checking every tablet, but the force/weight relationship is not sufficiently stable to be used as means of validating the product.

The ideal system would be one by which every tablet is weighed and accepted or rejected according to the limits set, coupled with feedback to ensure that the mean weight stays as near as possible to the required value. Unfortunately, with tablet machines capable of operating at speeds up to 13 000 tablets per minute, equipment to weigh at these speeds has still to be developed. The best compromise to this ideal system is to combine the force measuring and weighing systems to give a complete control of tablet weights. A system such as this is manufactured by Manesty.

The combined system operates in two ways. The force measuring system is used to check each tablet, and any tablet producing a suspiciously high or low force signal is automatically rejected. These can be recovered. The mean force is calculated and used to adjust the mean tablet weight, maintaining it as near as possible to the required value. The reject mechanism which is used to discard individual tablets with suspect force signals is also used to select sample tablets. These are transferred to a balance, which is linked to a computer. The computer triggers the reject mechanism to take sample tablets sequentially from each station of the press. If the station selected for sampling produces a suspect tablet, the tablet is rejected and the computerized weighing system waits for the next revolution of the turret to take its sample. If the station required continues to produce reject tablets a warning is given. Assuming the tablets are sampled successfully, 20 tablets are weighed and the results compared with the specification entered into the computer memory for that product.

If the sample of 20 is accepted the mean weight and standard deviation are printed for batch records and the weighing system continues to weigh the next 20 tablets. This system ensures that all stations are sampled at regular intervals, and there is no possibility of tablets from any particular station not being included in the weighing, as could occur with a random selection if the sample of 20 is outside the limits of the product. The machine is stopped and the 20 individual weighings are printed out so that the reason for the stoppage can be ascertained. A further refinement is possible where all the production, between one weighing cycle and the next, is held in a quarantine bin until the weighing is completed, and is added to the batch of accepted tablets only if the result of the weighing is satisfactory. This refinement is usually thought to be unnecessary, particularly as every tablet has been checked by force and a sample weighed, and it is unlikely that reject tablets will reach the quarantine bin. However, for high value tablets it might be considered worthwhile.

The most interesting part of the system is the link between the two individual sections. This is achieved by the computer calculating the mean weight of each 20 tablets and comparing the result with the required mean. If there is a discrepancy a signal is sent to the force measuring system to adjust the force/weight relationship in such a way that the mean weight of the tablets is brought nearer to the required value. After a very small number of weighings the mean weight is maintained very close to the required value. In addition, any variations which occur in the force/weight relationship due to changes in the bulk density of the feed material are corrected as they occur. The whole system is, therefore, able to achieve the following:

1. Each tablet is checked for compression force, and suspect tablets are rejected.
2. Sample tablets are weighed and batch records produced.
3. Mean weight is kept as close as possible to the required mean weight.
4. Changes in the force/weight relationship are detected and corrected, or the machine is stopped and the reason reported.

This system will give the most accurate control of tablet weights on a high-speed tablet machine of any system yet devised.

However, for the application of computer control the computer required should provide the following facilities:

1. Data logging and process control of various parameters: tablet weight, hardness, thickness, etc.
2. Batch control is required with various formulations and sequences.
3. Optimization of batch control parameters.
4. Flexibility to change the process, size of system, etc.
5. Local operator interface in a hazardous area.
6. Interface to field devices suitably protected against hazardous area (intrinsically safe circuits, etc.).
7. Operator control/display within the control room via VDU/keyboard/printer devices.
8. Calculations and optimization of sampled and/or measured variables for analysis.
9. Printouts and logs required.
10. System size relatively small (150 loops approximately).

With these requirements in mind it is suggested that either a centralized or distributed control system would be suitable. The centralized system is preferable to the distributed system for the following reasons:

1. It is smaller and cheaper
2. The system input/output interface is similar as the distributed system requires special housings.
3. It allows the possibility to use an existing computer.

Most process control and data logging manufacturers use their own computers, and their software structure may not be comparable with the in-house computer software. However, it is possible to provide the interface necessary, but this would require detailed study.

There are some further possibilities for the use of a computer control system that may be considered:

1. A dual processor computer to provide back-up security for control and data storage.
2. Links to an in-house computer can be accommodated if a separate system is used. Data can be passed to the in-house system for processing or display if required.

Automatic Control of Tablet Weight

All the manufacturers of tablet equipment produce computer control systems for their rotary presses, and the most important parameter to control accurately is the individual weight of the tablets produced. The level of sophistication of these systems varies from manufacturer to manufacturer, and also the type of system required by the producer of tablets. The complexity will vary, depending on the level to which they wish their operators to be concerned with the production process.

One system used widely is described here, including its control and monitoring performance.

The Enclosed Loop System

A diagrammatic representation of the arrangement of the closed loop control system is shown in Figure 8.2. The main components of this system are the electric transducer which is mounted on the pressure roll axle, a charge amplifier and oscilloscope, UV recorder pulse discriminator, and a stepping motor. The pulses from the electric transducer and the pressure axle are fed into the charge amplifier where they are amplified into a proportional voltage. From there they are directed to the pulse discriminator. The pulses can also be displayed simultaneously on an ultraviolet recorder and a recording oscilloscope. The output from the discriminator is fed into the stepping motor and the pneumatic valve.

Selection of the Stepping Motor

The main criterion that led to the choice of a stepper as the actuator was because of the perfect match of its digital characteristics to the requirements of the present

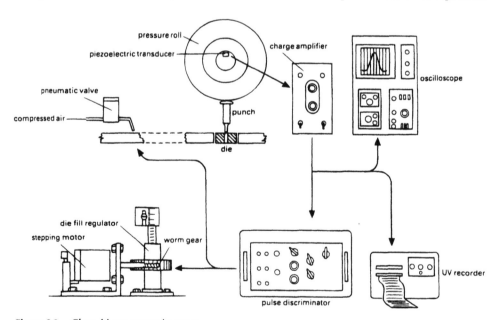

Figure 8.2   Closed loop control system

problem. Because of the nature of the problem (series of pulses), there is no need for analogue-to-digital, or vice versa, conversion which would need complicated and expensive equipment. Having chosen the kind of actuator, the next step was to determine its size by calculating the torque and moment of inertia of the load that would need to be rotated by the stepper.

The load to be controlled is the die fill regulator, consisting of a stainless steel sleeve and a threaded rod carrying a head-cam.

The bottom end of the sleeve is located in a cavity on the casing of the machine, while the head of the threaded rod can slide up or down in a cylindrical guide hole.

Rotation of the sleeve causes the head-cam of the rod to rise or fall, thus controlling the level of the lower punches at the filling point.

Pulse Discriminator

The next important step in the buildup of the closed loop system was the design of the difference element, i.e. where the input and output variables are compared and the error is computed. This difference element took the form of the pulse discriminator. If the amplitude of the pulses corresponding to the compacting force is outside a predetermined range (upper and lower adjustable limits), then a corrective pulse is sent to the stepping motor terminals, which moves the die fill regulator in the appropriate direction.

Rejection Mechanism

The ejection of the tablets from the dies takes place about a quarter of a revolution after they have been compressed. On the Manesty 'Betapress', for example, the 90 degree lag between compression and ejection corresponds to four punches. For this reason, the compression force signal, which determines whether or not the tablet is within the preset range, has to be delayed by four pulses, until that particular tablet is on the top surface of the die. At this moment, if it is within the range, then it is allowed to proceed to the normal take-off chute. But if it is outside the range (either below or above) then a blast of air from the pneumatic valve blows it off into the rejects chute.

The actual rejection system therefore consists of two components:

- The shift register, which receives one pulse for every over- or under-weight tablet at its information terminal.

- The valve is the second component of the rejection mechanism. By connecting the fourth output of the shift register to the pneumatic valve, the record of each tablet (whether or not it is within the range) is available at the moment when it is raised by the bottom punch onto the top of the die table. On other types of machine where the ejection point is more (or less) than four pressing stations away from the compaction point, the corresponding output of shift register can be selected.

The pneumatic valve was situated as near the ejection point as possible. This was to minimize the time taken for the jet of air to cover the distance from the output of the valve through a copper tube to the output of the nozzle. To ensure that the blast of air from the valve was directed and felt only by the reject tablet, without affecting the trajectory of the others, trial and error was necessary to position this nozzle.

The entire system operates at the speed of the clocking pulses, which is the machine speed. All delays, if set correctly for one machine speed, will be self-adjusting when the machine speed alters; in other words, the control system locks onto the tablet production frequency. It is easier to initiate the rejection motion of the tablets, in a radially outwards direction than vice versa, because of the centrifugal force acting on the tablet.

The reject tablets are directed upward into the take-off chute, which is divided into two narrower ones, by fixing a piece of Perspex sheet along the middle line of the existing chute.

Modifications of the Die Fill Regulator

It was found necessary to modify slightly the die fill regulator as follows:

- The number of threads per inch of the screw system was increased to make the adjustment of the punches finer. The male and female threads were machine ground to reduce the backlash, the friction, and striction to a minimum.

- The sleeve and rod assembly were changed, the outside diameter of the sleeve was reduced, and this enabled the diameter also to be reduced in proportion.

- The cylindrical periphery of the rod which comes in sliding contact with a groove in the wall of the machine was ground to a very smooth surface to reduce friction.

- The bottom end of the sleeve was also ground to a very smooth surface to rest on the thrust bearing.

Types of Equipment

The remainder of this chapter will describe those items of equipment that are used for the automatic control of tablet weight, and associated systems which can be incorporated into either an automatic computer monitoring system with a maximum of control from computers or can be incorporated into plants where there is only a very limited amount of control by those types of system.

Compaction Force Monitor (CFM)

This piece of equipment is the simplest in the range. It is designed to monitor both the mean compaction force and the force on individual tablets. The mean compaction force closely relates to the mean weight of the tablets, and indicates when changes in the mean tablet weight occur. It will also warn the operator when the change exceeds a predetermined value and can be set to automatically stop the machine when this value is exceeded.

136

Sentinel

This unit is designed to control the mean tablet weight. It contains all the features of the monitor. In addition the tablet weight adjustment is motorized and the unit compensates for any changes in mean tablet weight which occur. If any tablet is produced with an exceptionally high or low compaction force, a group of tablets containing this particular tablet is automatically rejected from the batch. If the level of rejection reaches five out of twenty tablets the machine is stopped. The unit also includes a counter which will show the total number of acceptable tablets produced over any given period.

There are a number of optional features available with the Sentinel.

1. Reject indication: This option consists of a two-digit display which indicates the number of any station which produces a reject tablet.

2. Individual tablet rejection: This option changes the reject system so that instead of a group of tablets containing the rejected one being eliminated from the batch, only the reject tablet is eliminated. It also includes the indication of the station number from which the tablet has been eliminated (option (1).

3. A recorder: A chart recorder can be incorporated which will produce a record of the mean tablet force/weight throughout the production run. It will also record if any tablets have been rejected.

4. A printer: This option provides a different type of record. Information on the batch, e.g. date, batch, method, operation, operating conditions, etc., can be entered at the start of the run. This will be printed for inclusion in batch records. The printer will record, at pre-set intervals, the mean force/weight over the period from the previous printout and the relative standard deviation of the force/weight over the period. It will also print the time, station number, and percentage increase or decrease in force from the mean value, of any tablet which is rejected. At the end of the run a summary of the batch information including the total number of accepted tablets, the number of rejects, the mean force/weight for the total batch, and the relative standard deviation.

Micro PW System

This unit represents a completely new approach to tablet machine control. The Sentinel offers an excellent means of machine control, but it cannot guarantee that the tablets produced have remained within specific weight limits. This also applies to any tablet press control system based on measuring the compaction force, irrespective of what the manufacturer's literature may imply. The relationship between tablet weight and compaction force holds good for most tabletting conditions, but there are many factors which can interfere with this relationship. (This relationship and the factors which affect it are discussed in Appendix 1). To overcome these possible errors the Micro PW combines automatic sampling and weighing of individual tablets, with a force measuring system. The two systems are combined in a way which gives absolute control of tablet weight.

The unit incorporates a Sentinel complete with individual tablet reject system. This is linked to the automatic sampling and weighing system by a microprocessor.

The microprocessor performs a number of functions. It is used to instruct the operator in the setting-up procedure for a particular product from information stored in its memory. It compares the results of the individual tablet weighings with standards set for the product, and accepts or rejects all tablets made since the previous weighing cycle. It also controls the sampling and weighing system and prints out the results for batch records. Its most important function is to relate the two systems so that the force measuring system is always related to actual tablet weight.

Sampling Unit

This equipment is derived from the Micro PW. It provides a relatively simple method of sampling individual tablets from a known station of tooling, when the machine is designed to operate in three different ways. It can sample tablets from any particular station of the machine; it can sample each station sequentially, or it can take one tablet from each station and one full revolution of the turret. The first and second methods of sampling can be operated either continually or on a set time interval, e.g. every 30 seconds or every 2 minutes, etc. The third method can be operated only on a set time interval because sampling one complete revolution of the turret continuously would result in the total output of the machine becoming samples.

Uses of Equipment

Compaction Force Monitor (CFM)

The monitor indicates to the operator any change in the operating conditions of the press. The meter on the front of the unit is set to zero when the press is producing tablets of the correct weight. The dial used to set the meter is calibrated so that it reads the action mean force being used to compress the tablets. Before the introduction of this equipment the actual force being used could only be estimated on machines fitted with an overload pressure gauge. This was done by reducing the overload pressure until the machines overload spring was activated. This then was a guide to the force being used. The use of the CFM enables an accurate force to be measured. It can be logged for a particular product and a normal operating range for that product established. Any batch of granules of that particular product requiring a substantially different compression force can be considered suspect and the reason investigated.

The meter on the CFM is scaled from -100 to $+100$ with the zero in the centre. When the unit has been set with the machine operating at the correct mean weight, movement of the meter needle indicates the percentage increase or decrease in the compaction force. The operator can quickly establish a relationship between the meter reading and tablet weight by noting the meter reading each time he/she takes a sample of tablets for weight checking. The meter reading on the CFM can be used as an instantaneous guide to the mean tablet weight being produced. Operators will also find that if the mean weight has deviated from the required value, moving the weight adjusting mechanism until the meter reads zero will then return the mean weight to the required value. This is much faster than making adjustments, rechecking the weight, and then making further adjustments until the correct weight is established.

Once a normal compaction force reading for a particular product has been established, it can be used to speed up the setting-up procedure each time the product is manufactured. The operator has to simply set the tablet thickness control to the correct setting for the product, set the adjustment on the CFM to the normal compaction force for the product, start the machine, and adjust the tablet weight mechanism until the meter reads zero. The tablets produced should then be at the correct mean weight. This method saves both time and material in setting-up the machine at the start of production.

If the operator finds that the meter reading is drifting further and further from the zero position but the tablet weights are remaining constant, he is alerted to the fact that conditions are gradually changing on the machine. If the drift is always to the plus side, one of the most likely causes is that the punches are becoming tight in their guides. The machine can then be stopped and the punch guides cleaned before any damage occurs.

The unit is also fitted with two set points which can be adjusted to any percentage increase and decrease in compaction force. If these limits are exceeded by 'the mean compaction force', then one of two lamps on the front of the unit will illuminate to indicate which limit has been exceeded. At the same time a relay within the unit changes state. This relay can be connected by means of a socket on the back of the unit to an alarm to warn the operator, or could be used to stop the machine. This feature can be very useful if an operator is looking after several machines. In this case, if the hopper runs empty or the material stops flowing into the feeder for any reason, the tablet will become light and the low set point will be exceeded. At this point the machine would be stopped before reject tablets have been deposited on top of a batch of satisfactory tablets causing the lot to be rejected.

The alarm or the link between the unit and the machine starter is not supplied as standard. This is because the equipment required will differ depending upon the type of machine to which it is to be fitted. Either or both can be fitted if the unit is purchased at the same time as a new machine, but any competent electrician should be able to make the appropriate connections where the unit is to be fitted to an existing machine.

One further feature of the unit is the provision of two co-axial sockets on the back of the unit. One is connected to the amplifier and the output from this socket can be fed to an oscilloscope so that the compaction wave form can be viewed. This can be an extremely useful research tool when investigating the compaction properties of different formulations or the effects of additives on a particular granulation. The wave form can also be recorded by a UV recorder or similar type of high-speed recorder. A normal pen recorder is not suitable except for very low tablet outputs, as the friction between the pen and the paper makes it too slow to record the wave form accurately.

The other co-axial socket gives an output in the form of a square wave. The horizontal portion of the wave is proportional to the peak force from each tablet. This output can also be viewed on an oscilloscope. Alternatively, it could also be fed to a pen recorder where the trace produced would indicate how the mean force/tablet weight had varied while the machine was operating. It would also show if the machine had been stopped at any time and for how long.

Pharmaceutical Production Facilities

Summary of Uses

1. Monitoring machine performance.
2. Measurement of actual compaction force being used.
3. Comparison of compaction properties of different batches of the sample product.
4. Monitoring mean tablet weight.
5. Faster setting-up procedure.
6. Indication of tight punches.
7. Alerts operator to stop machine for unacceptable operating conditions or in an emergency.
8. Investigation of compaction profiles, for use in formulation development of investigation of compaction problems.

Appendix 1   Correlation Between Compaction Force and Tablet Weight

The correlation between compaction force and tablet weight on a multi-station press will be valid only where the friction between the punches and their guides remains constant (i.e. the punches do not become tight); the overall length of the punches is constant; the tip geometry is constant; the die bores are uniform, and the pressure rolls are perfectly cylindrical and mounted centrally. The contribution of all these factors, with the exception of the friction between the punch and guide, to the total force measured on a stationary member of the machine, will be small. However, it cannot be ignored, and consequently there is no direct correlation between the compaction force and tablet weight for individual tablets.

The errors introduced by variation in the friction between the punches and their guides will, in most cases, consist in a gradual increase in measured compaction force for a given tablet weight. It is possible to monitor the frictional force for individual punches and to arrange for the machine to be stopped if it rises to a level which would invalidate the control settings being used. If the machine is kept clean and correctly lubricated the frictional force will remain constant and have no effect on any control system based on compaction force. In view of the fact that under correct operation conditions it will have no effect and even under adverse conditions it can be monitored and prevented from causing errors, it will be ignored in further consideration.

 ·The effect of the other factors means that it is possible for slightly different levels of compaction force to result from tablets having identical weights, and conversely identical levels of compaction force can result from tablets of slightly different weights. This variation will be limited because the variables causing it have a fixed range, i.e. punch lengths and die bores are not changing. Once the machine is set up with a set of tooling, the range of variation is fixed. Similarly, the effect of punch tip geometry and the degree of eccentricity of the pressure roll will be limited for a particular machine and set of tooling. Consequently, the correlation between the mean compaction force and the mean tablet weight is valid for each individual granulation bulk density, machine, and set of tooling.

If the compaction force/tablet weight relationship is plotted for a number of individual tablets having a range of weights (all from the same material and made on

140

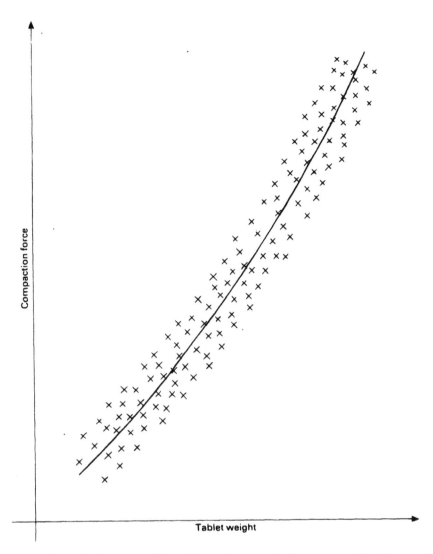

Figure 8.3 Relationship between tablet weight and compaction force on a multi-station machine. Each cross represents the compaction force and weight of an individual tablet

the same machine), a diagram similar to that shown in Figure 8.3 will result. The relationship between the mean tablet weight and the mean compaction force, over the range considered, can be obtained by plotting the best option through the points.

If the control system is to be used to reject individual tablets, where the relationship is not valid, the situation can be represented by Figure 8.4. This figure shows a plot of mean compaction force/mean tablet weight for a range of weights, slightly larger than the required control limits. The maximum acceptable tablet weight is indicated by the line 'H' and the minimum by the line 'L'. To ensure that all tablets above and below these limits are rejected, it is necessary to reject every tablet which results from a compaction force lower than 'A', or higher than 'B'. This means that

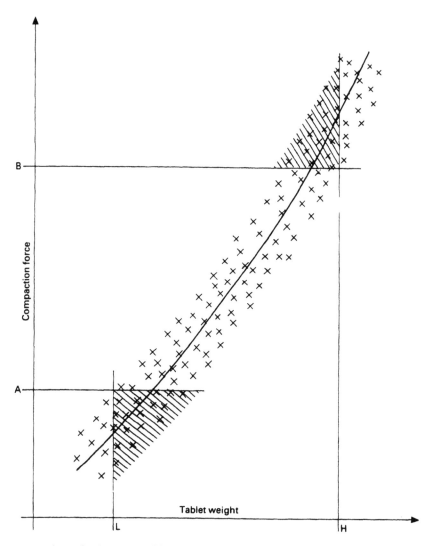

Figure 8.4 Relationship between tablet weight and compaction force on a multi-station machine where individual tablets are rejected. Each cross represents the compaction force and weight of an individual tablet

tablets represented by crosses, in the two shaded areas, which have acceptable weights, will be rejected. It has therefore been decided to refer to the tablets which are rejected by the force measuring system as SUSPECT, rather than reject tablets.

The quantity of good tablets rejected will depend upon several factors. It is possible, depending upon the range between the limits, that with a good granulation and a machine and tooling in good condition, that few, if any, tablets will be rejected. In this case the fact that some may be acceptable can be ignored. However, the importance of maintaining the machine and tooling in good condition, if this type of reject system is in use, cannot be overemphasized. Figure 8.5 shows the effect a machine and/or tooling in poor condition can have on the quantity of acceptable tablets which would be rejected. The shaded areas in Figure 8.5a, which represents a

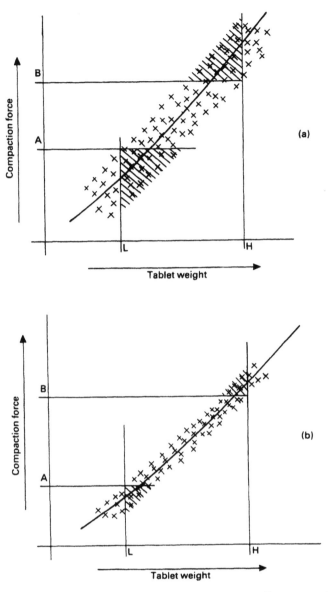

Figure 8.5 (a) Machine tooling in poor condition. (b) Good tooling, well maintained machinery. Each cross represents the compaction force and weight of an individual tablet

machine and tooling in poor condition, are much larger than those in Figure 8.5b, which represents good tooling in a well maintained machine. The amount of acceptable tablets rejected by a control system working under the conditions represented by Figure 8.5a, will, therefore, be much larger than that for Figure 8.5b.

If the tablets being produced are of low value, the fact that a few good tablets are returned for rework will not be significant. However, if the tablets are of high value and the quantity of suspect tablets rejected is significant, equipment is available which is capable of weighing and sorting all the suspect tablets at high speed so that the good tablets can be recovered.

NB. Figures 8.3, 8.4, and 8.5 are not taken from actual results but are diagrammatic, and the effects have been exaggerated to enable the situation to be more easily understood.

Tablet Sampling Device

This unit has been designed as a quality control or development tool. It enables individual tablets from particular stations to be sampled under normal operating conditions.

It uses a similar gate mechanism to that used in conjunction with the Sentinel or Micro PW. It can be used alone or in conjunction with the Sentinel. It cannot be used with the Micro PW as that unit already incorporates its own sampling system. Different parts must be supplied when it is to be used in conjunction with the Sentinel from those which are supplied when it is to be used alone.

The standard equipment consists of a control unit, a gate mechanism, a special take-off chute, and a pair of proximity detectors with a mounting bracket. The proximity detectors are used to identify the station which is to be sampled, and the mounting bracket will be specific to a particular type of machine. If they are to be fitted to an existing machine the stations must be numbered and a small hole drilled in the part of the turret containing the upper punch guides, between the last and the first station. On a new machine this operation would be carried out during manufacture. One detector counts the punches as they pass, and the other re-sets the count to one when it detects the hole after the last station. This part of the equipment is common to the options on the Sentinel for faulty station identification and individual tablet reject. If the sampling unit is being used in conjunction with a Sentinel with one of these options the signal can be shared, i.e. it is not necessary to fit two sets of detectors.

The signal from the detectors is fed to the control unit, and when the required station is detected the unit holds the information until that particular tablet reaches the take-off position. A signal is then sent to the gate mechanism to deflect the tablet into the sample section of the take-off chute.

The unit can be operated in various modes. It can be set to take sample tablets from one particular station. The station number is selected by a thumb wheel switch on the front of the unit. Alternatively, it can be set to sample sequentially from each station, or it can be set to take one sample from each station for one complete revolution of the turret.

9

Tablet Coating Systems

'Would you tell me, please,' said Alice, a little timidly, 'why you are painting those roses?'

Five and Seven said nothing, but looked at Two. Two began, in a low voice, 'Why, the fact is, you see Miss, this here ought to have been a red rose-tree, and we put a white one in by mistake; and if the Queen was to find it out, we should all have our heads cut off, you know?'

This would appear to be a very good reason for painting anything (film-coating is a painting process) and while the penalty for coating tablets the wrong colour is unlikely to be so extreme, the Queen (FDA, MCA, etc.) is likely to extract very costly and damaging retribution. No doubt 'heads would roll' metaphorically. So why are tablets coated? After all, it is a messy, complicated and expensive process.

'Look out now, Five! Don't go splashing paint over me like that!'

'I couldn't help it,' said Five, in a sulky tone. 'Seven jogged my elbow.'

It adds a degree of risk to the production process that could result in the whole batch being rejected. The costs in terms of space, personnel, equipment, quality control and validation are considerable.

The modern coating technique has developed over the years from the use of sugar to provide a pleasant taste and attractive appearance to tablets which were unpleasant to swallow due to their bitterness. There are, of course, many forms of coating which have a special function (such as enteric coating to delay the release of the drug until it reaches the intestine), but here the simple case will be examined. First, to answer the question 'Why are tablets coated?', a number of reasons can be suggested, some not quite so obvious as others:

- The core contains a substance which imparts a bitter taste in the mouth or has an unpleasant odour.
- The core contains a substance which is unstable in the presence of light and subject to atmospheric oxidation, i.e. a coating is added to improve stability.
- The core is pharmaceutically inelegant.

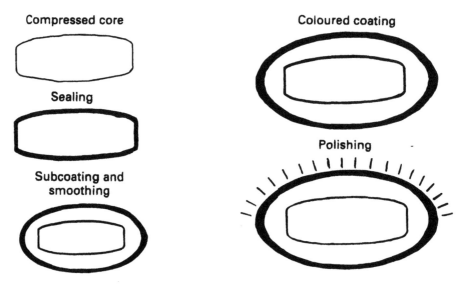

Figure 9.1 The stages in sugar coating

- The active substance is coloured and migrates easily to stain patients' clothes and hands.
- The coated tablet is packed on a high-speed packaging unit. The coating reduces friction and increases the production rate.
- To modify the drug release profile, e.g. enteric coating, sustained release coating, osmotic pumps, etc.
- Separates incompatible substances by using the coat to contain one of them or coating a pellet which is subsequently compressed into a core before coating.

This is not an extensive list but suggests several reasons for coating tablets. The process can be broken down into three main groups and one minor section:

- Sugar-coating
- Film-coating
- Particulate/pellet coating
- Compression coating.

There are several other historical coating processes such as pearl coating and pill coating which will not be discussed here.

With the exception of compression coating these processes rely on the continual application of sugar or a polymeric material to the tablet core as it rotates in a coating pan or is suspended in a fluidized cushion of air to build up micrometre thick layers. Detailed information relating to these materials can be found in *Pharmaceutical Coating Technology* (Cole *et al.*, 1995).

In the last 25 years tablet coating has undergone several fundamental changes. Coating of tablets and pills is one of the oldest techniques available to the pharmacist, and references can be traced as far back as 1838. The sugar-coating process was regarded as more of an art than a science, and its application and technology remained secret and in the hands of very few. Although a very elegant product was obtained its main disadvantage was the processing time which could last up to five days. Many modifications were advocated to improve the basic process such as air suspension techniques in a fluidized bed, the use of atomizing systems to spray on the sugar-coating, the use of aluminium lakes of dyes to improve the evenness of colour, and more efficient drying systems. However, the process remained complicated. Generally the sugar-coating process resulted in the weight of the tablet being doubled, but the use of spraying systems enabled this increase to be reduced dramatically. The two coating processes, sugar and film are schematically represented in Figures 9.1 and 9.2.

The first reference to tablet film-coating appeared in 1930, but it was not until 1954 that Abbott Laboratories produced the first commercially available film-coated tablet. This was made possible by the development of a wide variety of materials, for example the cellulose derivatives. One of the most important of these is hydroxypropyl methylcellulose which is prepared by the reaction of methyl chloride and propylene oxide with alkali cellulose. It is generally applied in solution in organic solvents at a concentration of between 2 and 4% w/v: the molecular weight fraction chosen gives a solution viscosity of 5×10^{-2} Ns/m^2 at these concentrations.

Many advantages can be cited for film-coating in place of the traditional sugar-coating process:

1. Reduction in processing time, savings in material cost and labour.
2. Only a small increase in the tablet weight.
3. Standardization of materials and processing techniques.
4. The use of non-aqueous coating solutions and suspensions.

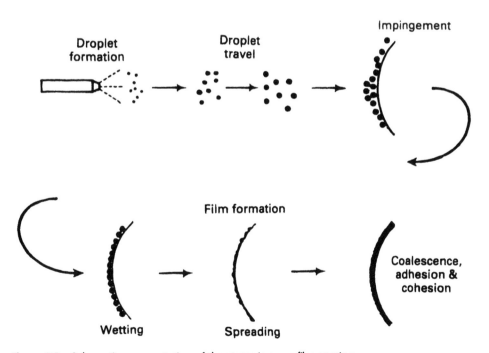

Figure 9.2 Schematic representation of the stages in spray film-coating

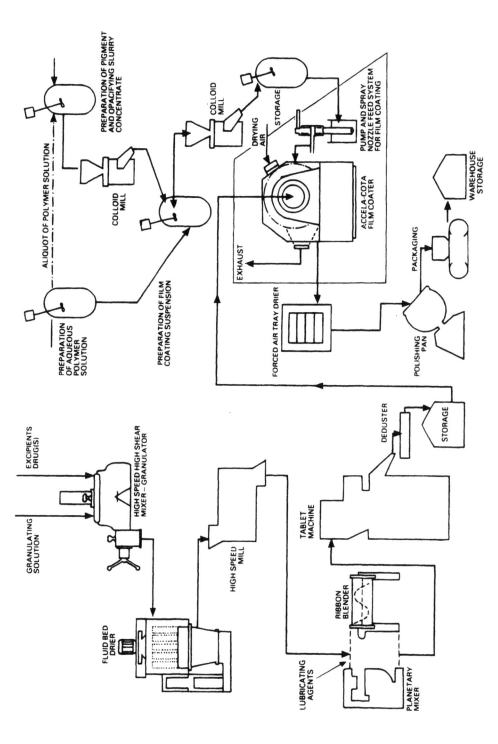

Figure 9.3 Flow diagram for the film-coating of pharmaceutical tablets

148

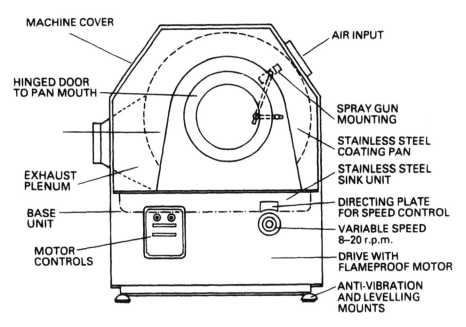

Figure 9.4 The Manesty Accelacota

5. The tablets could be engraved with a code and house logo which remained legible after coating. Many sugar-coated tablets were printed with a house symbol, name of product, or code after coating. This was a difficult and costly process which added nothing to the value of the product.

During the period 1954–1975 the lower molecular weight polymers of hydroxy-propyl methylcellulose with a solution viscosity of $3–15 \times 10^{-2}$ Ns/m² did not receive much attention because of the cheapness of organic solvents and the ease with which the coating could be applied. There was also a belief that the lower viscosity grades produced weaker films which would not meet the formulation requirement for stability and patient acceptability. However, there is now a trend towards aqueous film-coating for the following reasons:

1. The cost of organic solvents has escalated.
2. A number of regulatory authorities are considering banning chlorinated hydro-carbons altogether because of environmental pollution.
3. The development of improved coating pans and spraying systems has enabled these more difficult coating materials to be applied.
4. Flameproof equipment is not required, which reduces capital outlay, and a less hazardous working environment is provided for the operator.

Most of the early development work for aqueous film-coating concentrated on the use of existing conventional coating pans and tapered cylindrical pans such as the Pellegrini. This pan is open at front and rear, and the spray guns are mounted on an arm positioned through the front opening. The drying air and exhaust air are both fed in and extracted from the rear. The drying air is blown onto the surface of the tablets, but because of the power of the extraction fan most of the heat is lost with

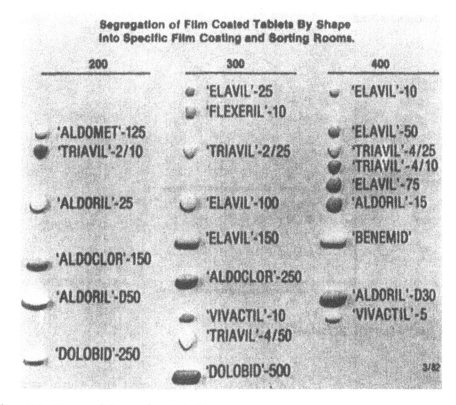

Figure 9.5 Types and shapes of coated tablets

the exhaust air. Very poor thermal contact results, and a poor coating finish is obtained. The perforated rotary coating pan which permits the drying air to be drawn concurrently with the spray through the tablet bed and pan wall during film-coating offers better heat and mass transfer and results in a more efficient coating process and a more elegantly finished product.

There are several companies which offer equipment of this type; the BWI Manesty Accelacota, the Driam Driacoater, and the Freund Hi-Coater are the three best known. There are significant differences between them.

Figure 9.3 is a flow diagram which illustrates the whole of the manufacturing process from mixing, granulating, compression, preparation of coating suspension, film-coating of the tablets, packaging, and storage of the product ready for sale. This book will be concerned only with the practical aspects of film-coating. The equipment used for this operation is outlined in Figure 9.3 and is shown diagrammatically in Figure 9.4. Figure 9.5 illustrates the types and shapes of tablets which can be coated using this equipment.

Mechanism of the Tablet Coating Process

Spray drying is widely used in the process industries to produce a range of heavy chemicals, food products, detergents, cosmetics, and pharmaceuticals, particularly antibiotics. Some of the theoretical and practical concepts of spray drying have been applied in this book to the aqueous film-coating process as applied to pharmaceutical tablets. One important difference between this process and conventional spray drying is that the atomized coating suspension is not completely dried by the time it strikes the tablets. Final drying takes place extremely rapidly, however, when the partly dried droplets come into contact with the tablet surface.

The tablet coating process as it occurs in the side-vented coating pan can be broken down for convenience into stages. The size, trajectory, and drying rate of the droplets as they move towards the tumbling bed of tablets also needs to be measured as a separate stage. This has led to a photographic investigation of the droplet's sizes and movement. The tablet bed itself is the location for the final drying; it is in some respects analogous to a packed bed humidifier, in that the air flows through the void space between the tablets in a mass transfer interaction with them, and it is important to know how closely the drying air will approach saturation in its passage through the bed.

These various stages are dealt with separately in what follows.

Atomization

This is one of the independent variables of the process. The ideal spray is one of small individual droplets of equal size. Heat and mass transfers and drying times are the same for all droplets in the spray, ensuring uniform dispersion on the tablets.

When correct atomization is achieved, all droplets arrive on the tablet surface in the same state, and in one revolution of the drum will have dried to increment the film-coating thickness without overwetting.

The invention of the mechanism theory which is applicable to commercial atomization is credited to Lord Rayleigh who in 1878 published a mathematical paper on the break-up of non-viscous liquid jets under laminar flow conditions. This was extended by Weber to include viscosity, surface tension, and liquid density effects (Weber and Angew, 1931). Later, Ohnesorge was credited with the following Reynolds number relationship: the tendency of the jet to disintegrate is expressed in terms of liquid viscosity (μ), density (ρ), surface tension (γ), and the jet size (d_n) (1936, 1937). The liquid break-up is therefore expressed by the magnitude of a dimensionless number Z' which is the ratio of the Weber number, We, ($V_j(\rho d_n/\gamma)^{1/2}$) to the Reynolds number:

$$Z' = \frac{\mu}{(\rho d_n \gamma)^{1/2}}$$

Although certain features are unique to particular types of atomizers, many of the detailed mechanisms of disintegration are common to most forms of atomizer. The most effective way of utilizing energy imparted to a liquid is to arrange that the liquid mass has as large a specific surface area as possible before it commences to break into drops. Thus the primary function of an atomizer is to produce thin liquid sheets. However, this is a simplification as all mechanisms tend to act simultaneously in commercial operations, influencing the spray characteristics to some extent. It was postulated that the nozzle should fulfil the following requirements:

• be capable of producing a droplet size spectrum of low mean diameter
• be able to handle a range of low viscous suspensions with a solids content of between 8 and 20% w/w

- be of simple construction
- have simple controls for altering the spray angle.

Two types of atomizer are considered. Traditionally for organic solvents both pneumatic and airless nozzles have been used for tablet film-coating, and both these types are examined. However, there are serious difficulties with the airless system for aqueous coating. In particular the higher spray velocity and the denser spray cone cause over-wetting, so that the tablets adhere to each other and to the walls of the coating pan. This method of atomizing the droplets has generally been discarded in favour of the two-fluid nozzle which uses air as the energy source to break up the liquid. This method satisfactorily produces a spray of droplets having a high surface-to-mass ratio. A high relative velocity between liquid and air must be generated so that the liquid is subjected to the optimum frictional conditions. These conditions are generated by expanding the air to high velocity before it contacts the liquid or by directing the air onto thin unstable liquid sheets formed by rotating the liquid within the nozzle. This provides very efficient and rapid formation of 20 μm mean diameter droplets. High and low viscosity liquids can be sprayed without difficulty. Because the flow rates and viscosity were low, rotation of the liquid within the nozzle is not essential for complete atomization.

In a laboratory unit used for experimental work, only one nozzle is required to provide the necessary spray rate. In the larger production units more than one of these nozzles is required. It was therefore necessary to use a type of nozzle with a controllable spread to ensure that the spray patterns do not overlap and cause overwetting of the tablets.

Nukizame and Tanasawa (1950) have shown that the mean spray droplet diameter D produced by pneumatic atomization follows the relationship:

$$D = \frac{A}{(\mu_{Rel}^2 \rho)\alpha} + B\left[\frac{W_{air}}{W_{liq}}\right]^{-\beta}$$

where u_{rel} is the relative velocity of air and liquid at the nozzle head and W_{air}/W_{liq} is the mass ratio of air to liquid. The exponents α and β are functions only of the nozzle design, whilst A and B are constants involving both nozzle design and liquid properties.

The mass ratio W_{air} to W_{liq} ranges from 0.1 to 10 and is one of the most important variables affecting droplet size. It has been reported that below 0.1 atomization deteriorates very rapidly and 10 is the limit for the effective ratio increase to create smaller sizes. Above 10, excess energy is expended without a marked decrease in the mean droplet size. It has also been reported that 5 μm droplets do not disintegrate into smaller sizes in the presence of high velocity air, but experimental sampling has shown particles as small as 1 μm to be present. From manufacturers' data for the nozzles commonly in use, a W_{air}/W_{liq} ratio of between 5 and 7.5, and an exit air velocity in excess of 300 m/s, it is predicted that droplets with a mean diameter of 20–30 μm would be obtained. The rationale for producing droplets of this size is to attempt to utilize the internal energy of the droplet as an aid to the evaporation of the droplet during its path from nozzle to tablet. Particles which are too small will be dried before striking the tablets and therefore the coat will not adhere to the tablet surface. As the latent heat of vaporization of water is so large, a combination of these energy sources can combine to dry the droplet completely immediately after striking the tablet.

Table 9.1   Droplet particle size spectrum

Particle size range μm	Cumulative per cent	Histogram per cent
below 5	10.8	10.8
6.6	32.7	21.9
9.4	51.0	18.2
13.0	63.0	12.8
19.0	79.4	15.5
27.0	94.7	15.2
38.0	99.1	4.4
53.0	99.1	0

Attempts to confirm these predictions have been made using two different approaches:

- photographic,
- impingement of particles onto microscope slides.

The photographic assessment of the droplet size and velocity distribution in an atomized spray presents no great problem when the size is 50 μm or greater, but below this, in-flight photography becomes more difficult, and attempts to establish a dynamic method were inconclusive. Most previous workers, including Groenweg *et al.* (1967) and Roth and Porterfield (1965) found that 10–20 μm presented the lower limit of size that could be photographed. Ranz and Marshall (1952), however, using high-speed cine, have produced shots of the thin sheets of liquid disintegrating into droplets.

The collection of droplets by impingement on to microscopic slides, Cole *et al.* (1980) clearly showed particles smaller than 5 μm. Similar results were obtained by a nozzle manufacturer (Schlick), using similar control parameters and measuring the particle size by means of a helium/neon laser and extracting the light energy from the droplet diffraction pattern. Some of these results are shown in Table 9.1.

Equipment

The following is a description of various units that are available for coating tablets, ranging from a simple home-made column with a capacity of 20–50 g to the large production machine capable of coating batches in excess of 1000 kg.

Coating columns – The original fluidized bed tablet coating column was designed by Wurster (1960), and it has been used extensively worldwide by a number of companies (e.g. Merck, Squibb) with high output requirements of film-coated tablets. It was particularly appropriate when solvent film-coated tablets were being produced. The solvent could be removed very quickly and a pharmaceutically elegant tablet produced. The construction of the Merck plants at Cramlington UK and Wilson USA used this principle to manufacture and coat tablets with a design capacity of 1000 million per annum. However, the development of the side-vented coating pan

(Accelacota, etc.) and the advent of aqueous film-coating processes has seen this method fall out of favour as the system of choice. It does have two advantages:

- the use of very small laboratory columns for initial formulation studies
- the coating of seed, pellets and very small particles where solvents are still used, then a totally enclosed system can be designed to recover all the solvent for disposal and protection of the operator.

Laboratory columns – Home made
One of the problems that exists in the earlier stages of formulation is the very small amounts of drug available. This column can be used to coat 10 g of pellets or tablets depending on the diameter of the column. The important criteria are:

- a pump capable of delivering 0.1 ml of coating consistently and accurately
- a controllable source of drying air
- a nozzle of 0.25 mm diameter.

Proprietary systems such as the Glatt can coat powders, granules, pellets, tablets, and capsules.

It can also be used for fluid bed drying and granulation with the necessary change parts.

Conventional Coating Pans

Various methods have been used to improve the coating characteristics in this type of pan which was originally designed for sugar coating:

- standard coating pan
- standard pan using the Glatt immersion – sword system
- standard pan using the immersion tube system.

Side Vented Perforated Cylindrical Coating Pans

The original type was the Pellegrini which was not perforated. However, with the development of the Eli Lilly pan (probably better known as the Accelacota) and the introduction of aqueous coating the side-vented pan is now the equipment of choice. Various forms of this pan are available. Some of the better known are:

- the Accelacota
- the Hi Coater
- the Driam Coater
- the Dumoulin Pan
- the Glatt Rapid Coater.

This type of coating equipment has largely taken over from the Wurster column for the film-coating of tablets owing to ease of application of aqueous coating formulations. The side-vented pan produces a more elegant product and can coat tablets of a higher friability than the column. One facility that still uses columns for aqueous coating is the MSD facility at Cramlington in the UK.

As the side-vented pan is now of major importance this chapter provides a background to its development and describes some of the problems that can arise and how they may be solved.

Description of Side-Vented Coating Pan

The side-vented cylindrical coating pan is discussed under three main headings: the initial development, the coating procedures found most successful for solvent/ aqueous coating, and the possible future development of the equipment.

Development

The original pan was developed by Eli Lilly & Co. in Indianapolis specifically for film-coating. Before its development Eli Lilly had been using both conventional pans and Wurster columns for film-coating, and the development of the side-vented pan was an attempt to combine the gentle action of the conventional pan with the speed of the Wurster column. It is important to remember that this work was specifically designed for solvent based film-coating. Some work was carried out on sugar-coating, but the technique gave very high drying rates resulting in a rough coating, and when the drying rate was reduced the improvement in coating time was lost.

The initial problems were associated with the low volume of air flow through the pan. However, by increasing the air flow, improvements were obtained for coating rates, appearance, and material recovery. A possible explanation for this was that the air being drawn through the Accelacota in the early experiments had a controlled relative humidity in the region of 30–35% whereas the air used routinely has relative humidities in the range of 50–70%.

A feature of the pan, which is now taken for granted, is its shape. Although it may not be immediately obvious, the two sides of the pan have different angles. This was the deliberate design feature to promote mixing, but subsequent work and experience with many products and batches found that the shape of the pan was not efficient enough to provide homogeneous mixing especially in the larger pan sizes.

Much of the very early work on the pan was concerned with mixing, and a wide range of devices were tested. These included a plough design and the final design which uses baffles.

The baffle design was originally thought to achieve the desired objective from the results obtained by performing enteric coating trials. Two batches of similar tablets were coated under identical conditions, one with baffles, one without. In the trial without baffles a coating of 18 mg per tablet was required before the tablets passed the BP enteric coating test. In the trial with baffles the tablets passed the test with only 10 mg of coating per tablet. This design did not suit all users and many requests were made for the design to be modified.

The change to aqueous coating systems has exposed the mixing limitations of the original baffle system especially where sustained release products are being coated and Figure 9.6 illustrates these baffles. However most equipment suppliers have developed unique baffle systems depending on the physical characteristics of the product that is to be coated.

BAFFLES IN PRESENT USE

BAFFLES

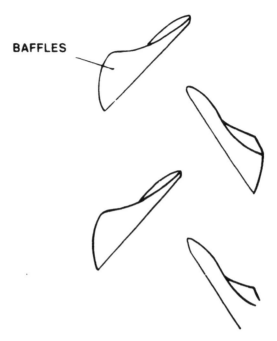

Figure 9.6   Original baffle design and configuration

Another variable which affects both the mixing and abrasion of the tablets is pan speed. Early trials were all performed with a batch size of 80–120 kg, and within a speed range of 7–21 rpm, based on data obtained from the original experimental work by Eli Lilly. Many users in the UK and Europe required only the lower end of this speed range, and it was considered that there could be some advantage in having an even lower minimum speed. In the USA at this time the most common pan speed used was 15 rpm. The speed range was therefore modified to 5–15 rpm. However, with the advent of aqueous coating there was a requirement for even lower speeds below 5 rpm. To provide this facility a simple variation speed unit with a wider range was developed. This used a double vari-speed pulley system where one pulley is spring-loaded and the side plates of the other are adjusted hydraulically.

This raises the question of what is the best speed to use for any particular coating conditions. Unfortunately there is no simple answer to this question. Low speeds are best for friable tablets, but for tablets which are not necessarily hard but non-friable, very much shorter coating times can be achieved by faster pan speeds.

The original plenum was designed small in order to achieve a uniform bed depth, and only covered that portion of the bed which was essentially uniform in depth. The air flows used were in the region of 1000 ft^3/min and could be passed through a small outlet without undue pressure drop. The plenum was designed to duct air from the bed of tablets either out of the back of the unit or into the base, where the exhaust fan could be located.

156

It was soon found that most of the early coating problems were associated with a lack of evaporative capacity, and experiments were carried out with much larger air flows. This resulted in a very high pressure drop through the small plenum. It was calculated that for an air flow of 2000–2500 ft^3/min the plenum outlet and ducting should be about 12″ diameter, and turning the air through 90 degrees in the plenum was undesirable. A new plenum was designed which covered a larger area of the pan which improved the appearance of the coated tablet and reduced processing times.

At this point the design of the American and European versions of the side-vented pans started to diverge. The American version used low air flow at a lower humidity, whereas the European version used higher flow at a higher humidity. This difference caused a certain amount of confusion to some international companies. They found that operating conditions which they believed to be identical were producing different results in Europe from those in America. This was explained by the difference in design of the equipment, but many companies now appreciate that where European plants are air conditioned, the air conditioning equipment has not been designed with sufficient excess capacity to allow low humidity air to be used.

Where the air flow has been increased to give satisfactory coating, it may be possible to reduce this increase by using heated air. The original design of the pan did not include a cover for the pan; air was simply drawn into the pan through the exposed perforations.

In any experiments in which heated air was used, the air was introduced through the mouth of the pan. This was found to be unsatisfactory for several reasons. The high volume of air had to be ducted through fairly small diameter piping (which meant that the air was travelling at high velocity) to permit access for the spray equipment, observation, and sampling. This caused disruption of the spray pattern and resulted in uneven coating. It was difficult to direct the air away from the sprays, and the warm air crossing the spray path caused spray drying of coating solution resulting in a loss of material. It was, therefore, reasoned that if air was brought into the pan through the perforations at 180 degrees to the plenum, it would be travelling in the same direction as the spray and would tend to carry it onto the tablets.

An enclosure was built around the pan and the air was introduced opposite the plenum. This was an improvement and it was assumed that the air that was entering the enclosure was travelling through the perforations, across the pan, and carrying the spray onto the tablets as intended. However, when this was checked by carrying out a smoke test, it was found that the perforations restricted the air flow into the pan and the whole case filled with hot air. The air was then drawn into the pan fairly evenly all round except for the area of the plenum. This meant that the air velocity in the pan was low and achieved the object of not disrupting the spray pattern.

The results of this work were interesting, in that, instead of reducing the amount of air drawn through the pan, it was found that better results were achieved, i.e. better quality of coating and shorter coating times. Consequently, heated inlet air very quickly became a standard feature of the European models.

Solvent Based Film-Coating

Solvent based film-coating was the system of choice, but this has now given way to the use of aqueous coating formulations. The most commonly used film former is

157

hydroxypropyl methylcellulose (HPMC). The low viscosity grades (5–65 cps) are used together with a plasticizer and colour. Typically, these can be sprayed from a solvent system consisting of 60% methylene chloride and 40% ethanol.

Coating formulation:

HPMC (15 cps)	300 g
Plasticizer (propylene glycol)	50 g
Colour (usually a 'lake')	qs
Methylene chloride	6.0 L
Industrial methylated spirits (ethanol)	4.5 L

Typical operating conditions are:

Exhaust air flow rate	140–160 m^{-3}
Temperature	Approx. 20–25°C
Inlet air flow rate	130–150 m^{-3}
Temperature	50–65°C
Relative humidity	40–60% RH
Quantity	Approximately 0.4 litres of coating formulation per kg of tablets is required for minimum coverage
Batch size	150 kg
Pan speed	6–7 rpm
Spray type	Airless
Spray guns	Two
Nozzle size	0.4 mm orifice/60% spray angle spray rate 1000 ml/minute upwards (dependent upon tablets and evaporative capacity available)
Liquid pressure	50–80 kg/cm^2 (dependent upon solution viscosity)
Coating time	Between 40 mins and 1 hour.

Experience has shown that airless sprays give the best results. Adequate coverage of the tablet bed can be obtained by using 4 guns in a 300 kg capacity pan; 2 guns in a 150 kg capacity model, and 1 gun in a 75 kg and 15 kg capacity model. Normally the best pharmaceutically elegant coating is obtained by spraying continuously. However, in the model 15 the minimum spray rate from one standard spray gun would cause over-wetting unless a specially designed spray nozzle is used. To overcome this problem an intermittent spray cycle can be used.

Various types of nozzle have been used. A 1.0 to 2.0 mm diameter orifice nozzle is fairly standard, and the usual reason for varying it is to prevent over-wetting or spray drying. Problems can arise when coating a small tablet which gives a very dense bed, with very small pore sizes and combined with fine intagliations on the tablets. The spray conditions quoted previously would have a tendency for infill of the lettering. Generally this is caused by the dense bed creating resistance to air flow and circulation and thereby reducing the evaporative capacity.

One method of preventing this is to reduce the droplet size. Increasing the liquid pressure may achieve the desired effect but may also increase the spray rate leading to over-wetting. Alternatively the solution may be diluted, thus increasing the spray rate by reducing the surface tension. When the spray rate is reduced, the degree of atomization will also decrease, creating a percentage of large droplets which may cause the original problem to return. In addition, the increased coating time could result in the coating wearing off the tablets almost as quickly as it is applied. The best solution is to reduce the nozzle size and maintain the liquid pressure at its original setting.

Generally, where coating bridges the engraving on tablets, and where it is not due to special characteristics of the process or tablets, one of the following reasons may be the cause:

- the spray rate is too high
- the droplet size is too large
- the spray guns are positioned too near the tablet bed.

Alternative Film-Coatings

Apart from HPMC based coatings, the side-vented pan has been used successfully for a wide range of solvent based film-coatings containing materials such as cellulose acetate phthalate (CAP), hydroxypropyl cellulose (Klucel), acrylic resins (Eudragit), shellac, and gum colophony.

CAP at a 1% w/w concentration coats particularly well from a solvent system consisting of 73% methylene chloride, 23% acetone, and 3% ethanol. Many variants are possible, and most companies using CAP coating have their own preferred solvent formulation. Experimental coating trials have shown that CAP coatings do not require hot air, and with certain solvent systems the use of hot air can prove to be a disadvantage.

150 kg batch of tablets
Coating formulation

CAP	5.4 kg
Diethyl phthalate	1.35 kg
Methylene chloride	54 L
Acetone	19 L
Industrial methylated spirits	2.7 L

Typical coating conditions for CAP are:

Pan speed	6–7 rpm
Spray type	Airless
Spray guns	Two
Nozzle size	0.4 mm orifice/60° angle
Spray rate	1250 ml/minute
Liquid pressure	54–82 kg cm^{-2} (as required to obtain spray rate)
Exhaust air	140–160 m^{-3}
Inlet air	130–150 m^{-3} 20°C (ambient) 40–60% relative humidity
Coating time	1 hour.

Sugar-Coating

Previous attempts to sugar-coat during the development of the side-vented pan were unsuccessful, largely because of the very rough coating produced by the high drying rates. Even when room temperature air, with the higher humidities common in Europe, was used for drying, the coating was still too uneven.

After a considerable amount of experimental work, a system of dosing, rolling, and drying produced satisfactory results comparable to that achievable in the conventional coating pan process. The drying air was shut off immediately after the application of sugar syrup was started, and by keeping the drying air shut off for a period after dosing was completed. During this period the sugar syrup spread by tablet to tablet contact until the whole batch was evenly covered and the thin film of syrup just started to dry. At this point the drying air was switched on and the maximum drying rate used, leaving a smooth coat.

The air flow was stopped during the dosing and rolling period by closing a damper in both the inlet and exhaust ducts. Stopping the fan was found to be unsatisfactory for two reasons:

- it took too long for the fan to slow down and stop
- even when the fan was stopped there was sufficient natural draught through the pan and ducting to cause some drying to occur, resulting in a slightly uneven coating.

Coating systems such as these are used with conventional pans. Here the drying time represents only about 30% of the total cycle time and consequently, if the side-vented pan could reduce the drying time by as much as 50%, it represents only a 15% saving in the total cycle time.

Advantages of the side-vented pan for sugar-coating:

- Semi-automatic operation
  The side-vented pan can be adapted quite readily to semi-automatic operation. An excellent example of this is in the sugar-coating department of Ayerst Laboratories at Rouses Point, New York.

- Dust free operation
  It provides a dust free operation, which virtually eliminates tablets with 'pimples' on them.

- Flexible in operation
  The operating conditions are very flexible, and can apply to a wide range of different coating operations.
  In extreme cases it can be used for removing sugar coats with blemishes. The sink can be filled with water and the tablets gently tumbled through the water until sufficient coating has been removed. At this stage the sink is emptied and the tablets rapidly dried to prevent damage to the core.

- Improved stability
  It has been reported that stability has been enhanced for some products coated in the side-vented pan compared with the same products coated in conventional pans. This may be due to a reduction in coating cycles and penetration of water into the core.

- Low temperature operation

This is a distinct advantage where heat sensitive tablets are to be coated, as ambient air can be used to dry the coating. This is also applicable to the confectionery industry in the coating of chocolate centres.

- Smooth coating
  An extremely smooth coating is achieved, even from the very early stages of coating. Excellent results have been obtained with a coating of only 50% of the core weight. Less could be used if it were really necessary. This is of little interest to companies wishing to copy existing products but can be attractive where new products are developed using sugar coating.

- Evenness of coating
  At high speeds the sugar-coating tends to follow the shape of the tablet. Normal or shallow concave tablets can retain their quite sharp corners after coating, even though the coating is smooth. If this is undesirable, a more rounded appearance can be obtained by extending the rolling time. At this stage the syrup is in a plastic state and will mould round the edges of the tablet.

Aqueous Based Film-Coating

The reasons for the introduction of aqueous based film coating have been highlighted in the introduction, and will not be repeated here.

The big disadvantage of using water as the solvent for film coating is the fact that it has a much higher latent heat of evaporation than most of the organic solvents (approximately three times that of ethanol: 539 K cal/kg cf. 204 K cal/kg).

The slower rate of evaporation gives rise to the possibility of water penetrating the surface of the tablet, which could result in either physical degradation of the tablet or deterioration of the active ingredient. The side-vented pan with its very high drying capacity makes it an obvious choice for aqueous coating. During the last ten years techniques have been developed to such an extent that even extremely water sensitive tablets have been coated without the penetration of water affecting the materials which comprise the core. Examples are aspirin, methyldopa, and vitamin C.

In developing successful aqueous film-coating there are a number of problems to be overcome. Most of the problems originate from the low evaporation rate of the water. This rate can be increased by increasing the air flow and/or its temperature. Increasing the air flow is not a simple solution. Time is required for the heat to be transferred to the aqueous phase and evaporation to take place, so that if the air flow is too high, heat is lost. In addition, increased air flow results in an increased pressure drop, through the bed of tablets and the associated ducting, and thus requires more power. The costs of higher air flow, additional power required and the heat losses, can increase without a significant reduction in the coating time. Consequently, there is an optimum air flow for each size of pan. Increasing the temperature of the air also helps, but this is possible only if the tablets are not heat sensitive. Results of many experiments have shown that the optimum inlet air temperature for most tablets lies in the range 50–80°C which maintains a bed temperature of approximately 40–45°C.

For an aqueous coating process the tablets must tumble in the pan for a longer period than would be necessary for a solvent based process. The most critical part of the process is at the start when the tablets are tumbling without the protection of a coating. At this stage, tablets which are friable may suffer erosion of the surface and edges. It should be remembered that the pan is acting in a way very similar to a large ball mill. This is also a difficult time to apply the coating quickly as any excess liquid on the surface of the tablet will penetrate the tablet core, and over-wetting can cause tablets to adhere to the pan walls and each other. The first stage of the coating process must, therefore, be a compromise between the need to apply the coating quickly and evenly to protect the tablet from abrasion, and slowly, to prevent water penetration and over-wetting.

If the tablets are relatively hard, i.e. non-friable or are not heat sensitive, then the coating process becomes relatively easy. Unfortunately, it is not always possible to produce tablets which will give the desired therapeutic effect and have the ideal physical properties. This was one of the major reasons for reducing the pan speed. While this helps to reduce abrasion it also reduces the frequency with which the tablets pass through the spray zone and increases the time they remain in that area. This, in turn, increases the possibility of over-wetting the tablets, resulting in uneven coating and an inelegant tablet.

The use of the baffles certainly reduces this problem, but it is essential that the action is gentle and that it provides homogeneous mixing even at low pan speeds. The object is to keep the tablets turning as they pass through the spray zone, thus preventing over-wetting. Baffles also prevent the formation of a very slow moving core of tablets in the centre of the bed, which have a tendency to receive less coating than the tablets travelling through the rest of the batch. It was found that the use of the baffles greatly assists in the coating of tablets which have a higher than normal friability. To minimize abrasion it is necessary for the baffle to be fully covered by the tablets. It is equally important to ensure that the pan is not overloaded as this tends to reduce the efficiency of the baffles.

Development of a Spraying System

The high degree of atomization which can be achieved from airless spray guns makes them the ideal choice for applying film coatings containing organic solvents. The very small droplet size gives a large surface area per unit volume, allowing a high rate of evaporation, and produces a very even coating. It was, therefore, considered that they might also be suitable for aqueous coating. However, for aqueous coating the spray rate is much lower, and to reduce the spray rate by using an airless spray gun requires either a lower liquid pressure or a smaller nozzle diameter. Reducing the liquid pressure leads to poor atomization and an uneven droplet size, and while reducing the nozzle size to a 0.07 mm diameter orifice obtained the required spray rate for aqueous coating, blocking of the nozzle became a serious problem. This nozzle size is suitable provided that soluble dyes are being used, but these are the exception rather than the rule as superior results for coloured films can be obtained by using insoluble aluminium lake pigments. With a nozzle size of 0.07 mm and incorporating these lake pigments into the coating formulations, repeated nozzle blocking occurs.

In the larger pans it was possible to reduce the number of spray guns, by increasing the liquid flow per gun and covering a larger portion of the tablet bed, but this resulted in uneven tablet to tablet coating. It was concluded, therefore, that airless sprays are not suited to aqueous coating.

Conventional air atomizing sprays can give excellent atomization of a liquid provided that the atomizing air volume is kept high and the liquid flow rate is low. The liquid feed to this type of spray gun is normally from a pressurized vessel (particularly for solvent solution), but it was found that this feed system did not provide a constant flow for aqueous coating. The small variations in flow rate resulted in variations in droplet size, and this manifested itself as an uneven coat or over-wetting. Many methods for obtaining a consistent liquid flow can be used for atomizing sprays in aqueous coating, and the Manesty C0-Tab Unit is one example. This was supplemented by a similar unit identified as the Spray-Tab unit and designed specifically for aqueous coating processes. These units were designed to achieve consistent flow of coating solution suspension, using a recirculating system. A pump was used to circulate the solution from a storage vessel to the manifold supporting the spray guns and back to the vessel. Even with a double acting piston pump two pressure regulators were required to even out pulses in the flow. The pressure in the circulating system and the rate of circulation were controlled by means of a needle valve in the return line. This maintained consistent pressure and flow even when the guns are repeatedly switched on and off. A fail-safe device was also incorporated into the unit to shut off the liquid flow to the guns if the pressure of the atomizing air falls below the level at which satisfactory atomization was achieved. This system has given excellent results with a wide range of aqueous coating materials including methylcellulose, hydroxypropyl cellulose, hydroxypropyl methylcellulose, and the acrylic resin suspension, Eudragit L30D.

Method of Operation

The operating conditions for the spray are critical. Some drying of the droplets takes place as they pass from the spray gun to the tablets, resulting in a coating that has partly dried when it strikes the tablet. The liquid is more concentrated when it is deposited on the tablet surface, and this together with the rapid evaporation of the remaining liquid prevents absorption on the tablet surface. The position of the spray guns in the pan is not extremely critical, but some care must be taken to ensure their correct operation according to the conditions being used.

The angle of the spray and its direction must be considered. Generally it is advisable to spray as far up the bed as possible to retain the tablets on the surface of the bed for the maximum time where a high evaporation rate exists. Tablets sampled from immediately below the spray zone will feel slightly tacky, but samples taken from the bottom of the bed, just before the tablets contact the pan and mix with the remainder of the batch, should have lost this tackiness. However, they will not be completely dry at this stage. If the tablets are still tacky at the bottom of the bed several options are possible:

- reduce the spray rate
- increase the temperature and/or volume of drying air
- check the angle of the spray produced and the air pressure used for atomization.

Tablets in this condition when mixed together will result in coating from one tablet sticking to another, and on separation a small portion of the coating is pulled off one of the tablets (commonly known as 'picking'). If 'picking' is allowed to take place a very inferior quality of coating is obtained. Final drying of the tablets takes place as they mix with the rest of the tablets in the bed. To complete the drying process before they pass again through the spray zone it is advisable not to have the spray directed at the very top of the bed as this portion of the bed is the least dense and the fastest evaporation takes place in this area.

Excellent coating results have been found by setting the direction of the spray approximately one third of the way down the bed. The distance of the spray nozzle from the bed will vary slightly with the materials being used and the operating conditions, but is usually between 250–350 mm from the surface.

If the guns are too near the tablet bed over-wetting may occur, and if they are too far away or if the temperature of the drying air is too high, spray drying occurs. This results in a loss of coating material through the exhaust system, blocked filters, and a build-up of material in the pan, plenum, and ductwork. When the correct conditions are used the recovery in terms of weight of coating deposited on the tablets to weight of coating material used, can be very high. Eli Lilly has claimed recovery as high as 99%, and while this figure has undoubtedly been obtained, more usual results are in the region of 95%. Spray drying of the solution can occur if the correct conditions are not used, and this emphasizes the need for spray guns to produce a very even droplet size. When a wide range of droplet sizes exist it will be difficult to establish optimum operating conditions.

Most of the process parameters are interrelated, and changes to improve one will affect at least one other. Fluctuating conditions will inevitably result in a poorer quality of coating. It is, therefore, essential to use equipment which will not only provide flexible operating conditions but must be capable of maintaining constant conditions, once they have been optimized.

Operating conditions for aqueous based coating

Coating formulation	% by weight
Hydroxypropyl methylcellulose (15 cps)	5.0
Polyethylene glycol (6000)	1.0
Colour (as solid matter)	1.25
Water	92.75

A typical set of operating conditions are given below:

Atomizing air pressure	5.0 kg/cm^2
Exhaust air flow rate	140–160 m^{-3} (see Note 1)
Temperature	38°C–42°C
Inlet air flow rate	130–150 m^{-3} (see Note 1)
Temperature	65°C–70°C
Relative humidity	41–60% at 15°C
Quantity	28 litres of solution
Batch size	150 kg
Pan speed	3.6 rpm
Spray type	Air atomizing
Spray guns	Three

Spray rate 100 ml/min/gun (dependent on
tablet shape and size)
Nozzle pressure 2.7 kg/cm^2
Coating time 90 minutes.

Note 1: Volume of air flow will depend on size of pan and batch size.

Future Development

Thermal efficiency

One area which is being investigated is the thermal efficiency of the coating system as a whole. It is common practice in the USA for units to be purchased without ancillary equipment. The user normally designs and instals his own exhaust and inlet air systems. In Europe and most other areas there is a demand for a complete package. Ideally, this should be a custom built system, each installation having its own design criteria. However, users in general have not attempted to optimize conditions, and once empirical operating conditions have been set they are then maintained without further investigation. Unfortunately, this has resulted in details of operating conditions, which are far from optimum, being circulated in the industry. The Accelacota in particular has been criticized on the grounds of poor thermal efficiency on the basis of this information. An examination of the results from several units from different manufacturers shows that coating cycles are very similar, some use longer coating periods but with lower air volumes while others increase the air volume to reduce the cycle. The cost of operating the process utilities is very similar.

Energy considerations

A comparison of the energy requirements for heating the drying air in a Model 350 Accelacota and a 450 mm diameter fluidized bed film-coating column suggested that the Accelacota was excessively inefficient in its use of the hot air for drying purposes. The following analysis (Table 9.2) does tend to superficially support this criticism, but the value of the hot air content of the exhausted air is negligible when compared to the selling price of the product. Also it does not illustrate the improvement in product quality achieved by using the Accelacota.

We shall consider the cost of coating 1 000 000 tablets (344 kg) of product in the Accelacota where C_p is the specific heat of air at constant pressure (kJ.kgK), T is its temperature (K), ΔT is the rise in temperature (K), G is the flow of air (kg/h), and Q is the quantity of heat (kJ/h), tablet bed temperature is 40°C, and the air temperature is assumed to be the same on leaving the pan before it passes through the exhaust fan.

The exhaust fan has a power of 20 HP or equivalent to 12 800 × 4.2 kJ = 53 760 kJ; 85% of this energy is transferred in the form of heat to the exhaust air, i.e. 11 000 × 4.2 = 46 200 kJ.

Table 9.2 Energy balance: comparison of a fluidized bed coating column and a model 350 Accelacota

	Model 350 Accelacota	450 mm column
Drying air volume m^3/h	8500	2000
Drying air mass kg/h	9010	2120
Inlet air temperature K	333	344
Outlet air temperature K	313	326
Temperature drop K	20	18
Spray rate kg/h	23	13.2
Energy drop kJ/h	$9010 \times 0.25 \times 20 \times 4.2$ $= 189 \times 10^3$	$2120 \times 0.25 \times 18 \times 4.2$ $= 40.1 \times 10^3$
Batch load: kg/h	140	50
Energy to evaporate water kJ/h	$23 \times 620 \times 4.2$ $= 59\,900$	$13.2 \times 620 \times 4.2$ $= 34\,372$
Energy loss kJ/h	$(189 - 59.9) \times 10^3$ $= 129.1 \times 10^3$	$(401 - 34.3) \times 10^3$ $= 5.8 \times 10^3$
Energy loss kJ/kg tablets	9.2×10^2	1.2×10^2

To calculate the temperature rise (ΔT) of the exhaust air due to this energy:

$Q = 46\,200$ kJ $\qquad\qquad G = 9010$ kg/h

$C_p = 1.05$ kJ/kgK $\qquad\qquad T = ?$

$$Q = GC_p \Delta T \quad \Delta T = \frac{Q}{GC_p} = \frac{46\,200}{9010 \times 1.05} = 4.8 \text{ K}$$

$\Delta T = 4.8$ K

This value is required to determine the heat content of the air being exhausted to the atmosphere.

If the temperature of the incoming air is 333 K and the ambient temperature is 293 K then the temperature rise is 40 K.

Total energy required: $Q = GC_p \, \Delta T$

$$= 9010 \times 0.25 \times 4.2 \times 40$$

$$Q = 38 \times 10^4 \text{ kJ.h}$$

To this is added the energy transferred from the exhaust fan.

Processing time = 2.5 hours

Total energy $= 95 \times 10^4 + 11.6 \times 10^4$ kJ

$= 106.6 \times 10^4$ kJ

Assuming that the cost of 1 kilowatt hour (kWh) of electricity is approximately £0.03 (1 kWh = 3.6×10^3 kJ),

Total cost $= \dfrac{106.6 \times 10^4}{3.6 \times 10^3} \times 0.03 = $ £8.88.

The cost of coating one batch of tablets in the Accelacota is approximately £9, assuming 100% utilization of the heat content of the drying air. This is 0.9 p per 1000 tablets.

The energy lost is the difference between the energy content of the air at exhaust temperature and the air at ambient temperature. The process time is 2.5 hours.

$$Q = \frac{G \times C_p \times (317.8 - 293) \times 2.5 \times 4.2}{3.6 \times 10^3} = \frac{9010 \times 0.25 \times 4.2 \times 24.8 \times 2.5}{3.6 \times 10^3}$$

$$= 163 \text{ kWh}$$

$$\text{Cost} = 163 \times 0.03$$

Total cost of lost energy = £4.9.

This is equivalent to a loss of 0.49 p per 1000 tablets or 55% of the total energy costing cost.

From this it can be seen that there is very little potential saving in the cost of operating the Accelacota at temperatures in excess of the minimum required. However, the use of higher volumes of air results in a faster process and a more elegant product. Saving in process operator time is much more cost effective than reducing the temperature and volume of the drying air. There are, however, reasons for ensuring that excessively high drying air temperatures are not used. These are:

1. Spray drying of the film coat before it is deposited on the tablet.
2. Excessive loss of coating materials due to spray drying.
3. Contamination of the exhaust system with large quantities of film coating due to the spray drying effect, resulting in increased maintenance costs.
4. Excess coating material deposited on the pan walls.

Coating Materials

This is an area which is consistently expanding as new materials become available. The use of starch or modified starches as possible coating materials is being investigated. These materials have been shown to offer advantages in terms of cost and to some extent speed of application.

Treatment of Exhaust Gases from Film-Coating Processes

No coating process is 100% efficient in terms of the amount of coating used to the amount actually deposited on the tablets. Some losses will always occur. The efficiency of the process is very difficult to measure as the tablets themselves may lose weight by abrasion during the coating process. Weighing the tablets before and after coating will, therefore, not give the weight of coating deposited. Various workers have devised methods of measuring the amount of coating applied and have claimed efficiencies of 85–95% and even higher. Some of the lost material will be deposited on the pan and some will escape with the drying air.

Material can be lost from the tablets in two ways. If the tablets are not dedusted before they are put in the pan, the dust on them will be removed by the tumbling action and the exhaust air. If tablets are left rolling in the pan for any length of time, without sufficient coating being applied, then the frictional forces of the tablets being in contact with each other and the pan will result in some weight loss. This material will be removed with the exhaust air.

In addition to the particulate solids in the exhaust air, the solvent used to apply the coating will be present. For sugar coating or aqueous based film coating it is not necessary to remove the water from the exhaust gases, but where organic solvents are used they must be removed to prevent environmental pollution. If they can be recovered in a usable form, cost savings can be achieved which will offset the cost of plant used for their recovery.

The total quantity of heat used in the coating process is relatively small. Its value will depend upon the cost of the fuel used and the efficiency of the heating system. However, if some of this heat can be recovered in combination with solvent recovery it can further improve the economics of the total coating process.

Cyclones

The simplest and cheapest method which can be used to remove solid particles present in the exhaust gases from a coating process is the cyclone. Air enters this equipment tangentially at the top and is forced to spiral downwards into the bottom section. As the dust particles in the air have a much larger mass than the gas molecules the centrifugal force exerted on them is much larger and they are thrown to the wall of the cyclone. They pass down the wall of the cyclone with the air and are collected at the bottom. The gas then flows up the centre of the cyclone in a much smaller spiral and escapes at the top. The solid particles which collect at the bottom of the cyclone are continuously removed by means of a rotary valve or via an air lock with automatically operated flap valves.

This equipment has the advantage that it is cheap to build and instal, it has no moving parts, and therefore it requires little maintenance. Its main disadvantage is its efficiency. Cyclones are usually suitable for removing particles larger than 50 μm, and many particles smaller than this are present in coating exhaust gases.

For the particle to be removed from the air stream the centrifugal force on it must be greater than the drag of the air which tends to carry it away. To increase the centrifugal force the diameter of the cyclone must be reduced, and this will in turn increase the pressure drop across it and hence the power required to drive the air through.

Some manufacturers now produce high efficiency cyclones, and these are usually operated as a batch of small units to obtain the required capacity, without introducing an excessive pressure drop. These can be suitable for removing in excess of 95% of all particles larger than 5 μm.

It is difficult, with this type of equipment, to specify exactly what its performance will be unless trials are made, using the particular conditions under which they will be required to operate. Sometimes it is possible to obtain much higher efficiencies than those predicted by theoretical calculations. Particles can agglomerate in the cyclone, and by this means, a much larger percentage of the small particles may be removed.

Another problem is that the efficiency can be affected by changes in the air volume, and one of the greatest causes of loss of efficiency is leakage of air into the cyclone at the point where the solids are discharged. For good efficiency the discharge mechanism must be well maintained.

Fabric Filters

These are probably the most commonly used method of removing dust particles from air streams. The design of the units varies considerably from one manufacturer to another. The filter elements can either be in the form of bags or as candles. The main object of the design is to make the maximum surface area of cloth available for filtration in the minimum space but in such a way that the whole of the surface area of the cloth is exposed to the contaminated air.

A wide variety of types of fabric are available for these filters so that the most suitable degree of filtration can be selected. This type of filter is capable of 99% efficiency with particles down to submicrometre size. A common specification for cleaning exhaust gases from coating operations is 98% efficiency down to 5 μm. The performance of the filter varies very little with air flow rate; but the surface of the fabric will gradually become coated with the particles being removed from the air. As this happens, the pressure drop across the filter will increase. This will reduce the air flow through the coating pan, which will in turn affect the coating process. It is, therefore, necessary to monitor the air flow and control it to a constant level, to ensure consistent coating conditions.

Most filters of this type are fitted with some means of automatically cleaning the filter surface. This can either be by mechanical shaking or by blowing air back through the fabric in the opposite direction to remove the particles adhering to the surface. Obviously filtration cannot take place while the cleaning is proceeding. As it is also detrimental to the coating process to stop the air flow each time the filter is cleaned some method of maintaining the air flow during cleaning must be devised. One possible solution is by arranging the filter in a number of sections, e.g. four. In this case three sections would be filtering the air while the fourth was being cleaned, one section at a time being automatically shut down and the exhaust air from the coating process directed to the other three.

One fairly recent development which has improved the cleaning of filters is the introduction of a fabric which is coated on one side with a 'plastic' membrane. One surface of the cloth is coated with plastic which has a very small pore structure compared with that of conventional felt. This prevents the particles penetrating the surface of the felt, and the particles, therefore, have much less tendency to adhere to the surface of the filter. The filter is consequently much easier to clean. It also allows the filter to operate for long periods with a low pressure drop (i.e. near to new fabric conditions) than is possible with traditional filter cloths.

Fabrics have the advantage that they can be chosen to form the filter so that the required degree of filtration is obtained and are also resistant to attack by organic solvents if they are present in the exhaust gases. Their main disadvantage is their physical size.

Wet Scrubbers

A much simpler and cheaper piece of equipment than the cloth filter is the wet scrubber. This type of equipment takes many forms but it essentially consists of a

169

two stage process. Both stages take place in the one piece of equipment. In the first stage the exhaust air is mixed with water from the spray, so that the solid particles are wetted or more commonly they are captured by the drops of scrubbing liquid. The air is then turned through 180 degrees and passed into a much larger diameter chamber so that its velocity is slowed. In this stage the larger drops fall back and the smaller ones are removed by the mist eliminator.

The design of the first part is critical and differs considerably from one manufacturer to another. The object of this part is to get the maximum particle/water contact so that the maximum amount of solids is removed. If good contact is achieved then even a moderate proportion of submicrometre particles and over 90% of particles as small as 5 μm can be removed.

The design of the second mist-elimination stage is equally important, and again designs vary widely from one manufacturer to another. The danger in this part of the equipment is that any small droplets of water which do pass through the mist eliminator will be carrying some solid particles and therefore the efficiency of the unit will be reduced.

The main advantages of this type of equipment are:

- its small size – considerably less than that required for the equivalent cloth filter
- low maintenance costs
- low power consumption.

Its main disadvantage, if in fact it is a disadvantage, is disposing of the solution/slurry which collects in the base of the unit. It is usual to operate the unit on a closed system, i.e. the water is not continuously run to waste. If the water was continuously run to waste the cost of the water used could be quite considerable. After several batches of tablets have been coated it will be necessary to dispose of the solution/slurry from the tank, and as this may contain a certain amount of active material from the coated tablets it will not be acceptable to dispose of it in the normal drainage system. If scrubbers are used in conjunction with a sugar coating process, the dilute sugar solution which collects in the scrubber is an ideal medium for bacterial growth. It is, therefore, essential that it is cleaned out regularly. In fact scrubbers have been referred to as units which transfer an air pollution problem into one of water pollution.

A further point which should be referred to at this stage is that unlike the previous two methods, i.e. cyclones and cloth filters, the wet scrubber is also capable of removing part of any organic solvents which may be present in the exhaust gases. This applies particularly to water soluble solvents such as alcohol. This type of scrubber is not particularly suited to this application, and gas scrubbing for removal of solvents will be dealt with later.

Cyclone Scrubbers

One of the ways of improving the efficiency of a water washing system for solids removal is to combine the advantages of the cyclone with that of the simple scrubber. This type of equipment can also vary widely from one manufacturer to another. Depending upon the design, the water can be introduced at the top, bottom, or even on the central axis. The wet cyclone offers some advantages for certain types of gas cleaning, but it is doubtful if this type of equipment could offer any substantial advantages over a simple scrubber for the cleaning of exhaust gases from a tablet coating plant.

Electrostatic Precipitators

Another type of gas cleaning which is sometimes used is electrostatic precipitation. However, it is not known whether this type of equipment has been used for cleaning the exhaust gases from tablet coating. It is very good for removing small particles from gas streams. Electrostatic precipitators are capable of efficiencies as high as 97–98% for particles down to 0.05 μm and are suitable for very large gas volumes. However, their installation and running costs are much higher than for any of the other equipment described, and there could be dangers if inflammable solvents are being used.

Removal of Organic Solvents

Some organic solvents can be removed by washing the exhaust air. However, the type of wet scrubber already described is not designed to obtain the best results for the maximum removal of solvents. Other problems occur if solvents such as methylene chloride are being used. Methylene chloride will, to some extent, decompose to produce a dilute solution of hydrochloric acid, which is corrosive. Solutions containing methylene chloride may or may not be acceptable for disposal in local sewers. To overcome this problem one European pharmaceutical company has made the decision that all its organic solvent based coating will be carried out without the use of methylene chloride, i.e. it will be alcohol based. This means that the exhaust from the coating plant can be cleaned and the solvents removed by simply washing with water.

Gas Absorption Towers

For the maximum efficiency of solvent removal the objective of the equipment is to generate the maximum exposure of gas and liquid surfaces to each other. This can be done in three ways:

- The liquid can be broken up into a number of slow moving films which are dispersed through the gas.
- The liquid can be broken up into as small a droplet size as possible and dispersed into the gas steam.
- The gas can be broken up into small bubbles that are passed through the liquid.

As with all types of gas cleaning equipment the design varies widely from one manufacturer to another. Often two or all three of the above actions take place in different parts of the equipment.

One system which could be suitable for use with exhaust gases from a coating plant consists of two towers, a short, fairly large diameter tower and a much taller tower with a smaller diameter. In the first tower water is sprayed as very fine droplets to remove some of the solvent. In the second tower the gases flow upward through a series of trays against a downward flow of water.

So much has been published on the design of these absorption towers and there are so many possible designs for the internal structure that it is impossible to deal with all the variations. If this is considered an option then a specialist company should be consulted.

The main advantage of this type of cleaning process is that it is probably the cheapest means of efficiently removing solvents such as alcohol. Its disadvantage is that the solvent is lost as the solution is normally too dilute for recovery to be an economic proposition.

Carbon Absorption Systems

As the price of organic solvents is high it can form a significant portion of the total cost of the coating. There is obviously a case for trying to recover them. One possible method is by the use of activated carbon which can absorb the solvent vapours on its surface.

A carbon absorption plant consists of two or more towers each containing a bed of active carbon. The solid particles are first removed from the exhaust gases as these would tend to block the carbon bed. The gases are passed through the first tower or the first set of towers depending upon the total quantity of gases to be treated. The solvent molecules are then absorbed onto the carbon. This continues until a significant amount of solvent can be detected in the gases leaving the tower, i.e. the carbon bed has become saturated with solvent; here the gases are diverted to the second tower where absorption of the solvent continues. The first tower is then stripped of solvent by passing the steam through the carbon bed. The steam/solvent vapours are condensed and the solvent/water mixture collected. The carbon bed must then be dried and cooled ready for the exhaust gases to be passed through again. The size of the plant is to some extent governed by the time of the stripping, drying, and cooling stages, in other words how many towers or what size of tower is required to absorb the solvent vapour until the first unit can be brought back into use.

For tablet coating, which is frequently a batch process with a period between batches, it might be possible to use just one tower. If the times between batches are sufficient for stripping, drying, and cooling to take place or the tower has a capacity to absorb the solvent from several batches before stripping, then a one tower system would be suitable. This would obviously reduce the capital cost of the plant.

The major disadvantage of this system is that the collector contains a mixture of the solvent or solvents used and water. It is usually necessary to distil this mixture before the solvent can be reused, requiring additional plant and higher costs. If a solvent mixture has been used and this cannot be completely separated by distillation then analysis will be required and solvent will need to be added to return the mixture to the original proportions.

Various reports on the economics of operating this type of plant in connection with film-coating give different results. Early reports said that the value of the solvent recovered offset the operating costs of the plant, but the capital cost had to be considered as the cost of complying with anti-pollution legislation. More recent reports indicate that it is possible to obtain a return on the capital invested.

Condensation Systems

The alternative method of recovering the solvents is by condensation. A very interesting paper was published in *Die Pharmaceutical Industry* by Koblitz, Bergbauer-Ehrhardt of Sandoz Nurnberg, Germany, concerning this particular method. The system it is using, which at this stage is very much experimental, has many advantages over carbon absorbers. The exhaust gases leaving the coating pan are first filtered and then passed through an air to air heat exchanger. In this stage they are cooled from about $50/\pm45°C$ to about $+7$ or $+8°C$. The gases then pass into a condenser where they are cooled in two or three stages to $-30°C$. It can be shown from vapour pressure calculations that at this temperature 98% of the solvent vapour will be condensed. This liquid is collected and is in a form where it can be re-used without further treatment. The cold gases are then passed through the other side of the first exchanger, where they cool the air, leaving the coating plant before it enters the condenser and in turn are warmed by that air. The exhaust gases then pass to a reheater. The reheater and condenser are linked by a heat pump so that at least some of the heat removed is replaced.

The gases are then returned to the coating pan via a final heat exchanger which is used to bring them up to the required inlet temperature.

This type of plant will be expensive to instal but apart from preventing any pollution of the atmosphere, it recovers the solvent in a usable form and reduces the heat input thereby giving further savings in operating costs.

One difficulty which can occur with a system of this type is the control of any water which enters the system. If ethyl alcohol in the form of industrial methylated spirit is used as one of the solvents this will contain water. This water, as well as any entering the system from any other source, is likely to form ice on the heat exchanger surface of the condenser. This will reduce the effectiveness of the condenser and could hinder the operation.

There are various ways in which water can be removed, and if this is done the system probably offers one of the most cost effective ways of operating an organic solvent based film-coating process.

Five years ago it looked possible that aqueous based film-coating could take over completely and that organic solvent based coating could gradually die. However, quite a number of companies are reporting that they have at least one product which must be coated from an organic solvent based coating. Once the equipment is installed for solvent recovery, and if it is as economic to operate as this last system appears to be, then there is no reason why other products should not be coated from organic solvent based solutions. It will be interesting to see what the future holds.

References

COLE, G. C. *et al.*, *Pharmaceutical Coating Technology*, Taylor & Francis (1995).
RAYLEIGH, LORD, *Proc London Math. Soc.*, **10** (1878).

Pharmaceutical Production Facilities

WEBER, C. and ANGEW, Z., *Math. Mech.*, **11**, 136 (1931).

OHNESORGE, G. (a) *Z. Angew, Math. U. Math.*, **16**, 355 (1936). (b) *Zeitschrift des Vereins Deutscher Ingenieure* **81**, No. 16 (1937).

NUKIZAMA, S. and TANASAWA, Y., Translated by E. Hooc into English, Defence Research Board, Dept. of National Defence, Canada. 10.M9.47 (393) HQ 2-0-264-01 March (1950).

GROENWEG, J. *et al.*, *Brit. J. Appl. Phys.*, **18**, 1317 (1967).

ROTH, L. O. and PORTERFIELD, J. G., *Trans. Am. Soc. Agric. Engrs.*, **8**, 493 (1965).

RANZ, W. E. and MARSHALL, W. R. *Chemical Engineering Progress*, **48**, 3, March 1952.

COLE, G. C., NEALE, P. J., WILDE, J. S. and RIDGWAY, K., *J. Pharm. Pharmac.*, **32** (Suppl.) 93 (1980).

SCHLICK, G. COBURG, Germany, nozzle manufacturer.

WURSTER, D. E., *J. Apla. Sci. Ed.*, **49**, 82, (1960).

List of symbols

Parameter	Symbol	Units of measurement
Air flow	G	kg/h
Constant	A	Relationship between nozzle design and liquid to be sprayed.
Constant	B	Relationship between nozzle design and liquid to be sprayed.
Diameter of tablet	l	m
Density		kg/m
Dispersion factor	α	Function of nozzle design
Droplet diameter	D	m
Droplet size group		Function of nozzle design
Kilowatt hour	kWh	–
Hour	h	–
Horse power	HP	–
Liquid jet velocity	V_j	m/s
Nozzle orifice diameter	d_n	m
Mass	W	kg
Ratio of the Weber number (We) to the Reynolds number (Z)	$\mu(\rho d_n \gamma)^{1/2}$	–
Specific heat at constant pressure	C_p	J/kgK
Surface tension	γ	J/m
Tablet bed thickness	Z_s	m
Temperature	T_a, T_s, T_1, T_2	K (°C)
Temperature difference	ΔT	K (°C)
Total heat transferred	Q	kJ/kg
Velocity	u	m/s
Viscosity		Ns/m^2
Weber number (We)	$V_j[(\rho d_n/\gamma)^{1/2}]$	–

10

Capsule Filling Systems

*... the last time she saw them, they were trying to put the Dormouse into the
teapot.*

Capsules are solid dosage forms with hard or soft gelatin shells of various shapes
and sizes, usually containing a single drug substance and a number of excipients
formulated to provide a consistent fill weight and therapeutic relief when adminis-
tered to a patient. The soft shell gelatin capsule originated in France when Mothes,
a French pharmacist, was granted a patent in 1833. The first reference to the hard
shell two piece capsule was made in 1848 when Murdoch, a London patent agent,
was granted a patent. Hard shell capsules permit the choice of a single drug or
combination of drugs at the exact dosage level required by the patient who very
often finds swallowing a capsule easier than a tablet. The hard shell capsule is often
referred to as the dry filled capsule (DFC) but with the development of new formulation techniques and the improvement in the closing mechanism of the cap and body,
oils and pastes may now be filled. Considerable improvement in the equipment used
for this process has led to renewed interest, and Eli Lilly & Co. (the empty capsule
manufacturing and marketing operation of Eli Lilly has now been taken over by
Shenogi, the Japanese company) has held symposia on this topic devoted to the
presentation of papers on formulation and machine demonstrations. The first
British pharmaceutical product using this type of formulation was Colpermin.

Basically there are eight sizes of DFC in general use:

Capsule size	000	00	0	1	2	3	4	5
Volume ml	1.37	0.95	0.68	0.50	0.37	0.30	0.21	0.13

There is also the Capsugel Supro which is available in sizes A, B, C, D, and E and
volumes of 0.68 ml, 0.50 ml, 0.37 ml, 0.30 ml and 0.21 ml respectively. Special
machine change parts are required for Supro capsules. Size 5 are difficult to handle
and fill automatically and are generally used for toxicity and dosage studies for new
chemical entities. Size 000 and larger are used for veterinary products as they are
too large for patients to swallow.

Soft shell capsules usually contain liquids or paste. They can be of any shape or
size, but attempts to fill them with powders have not been very successful.

The development of the methods of filling powders into two piece gelatin capsules mainly took place in the United States, where some 80 patents were granted

between 1850 and 1900, in contrast to Britain, where only 15 patents were taken out in the same period, and only one dealt with the actual filling of capsules.

The stimulus in the USA originated from the company of H. Planten and Son, of New York, which commenced operations in 1836. Planten introduced the European process, and manufactured both the original Mothes capsules and two piece capsules filled with powder. When powder filled capsules first originated is not clear. Small hand fillers were used at this time, and the first automatic machine was not developed until the early twentieth century.

However, the pattern is beginning to change: in Western Europe, for instance, consumption of hard shell capsules rose annually by between 8 and 21% for the years 1971–80, while other solid dosage forms rose by only between 4 and 6%. Annual production is now in excess of 100 thousand million worldwide. It is apparent, therefore, that more formulations are being put into capsules, and because of their versatility more marketing teams are projecting their company image to the doctor and patient by using capsules. Some of the possible different physical combinations of products that can be filled into capsules are shown in Figure 10.1.

All the leading machine manufacturers sell equipment which will fill powder into capsules, though special dosage devices may be required to fill the other combinations shown. The range can be expanded considerably by changing the form of the pellets or tablets; they may, for example, be enteric-, sugar-, or compression-coated. The use of sustained release enteric coated pellets increases the possibilities still further.

Every filling machine must carry out four operations: rectify the capsules, separate the body from the cap, fill the body, and replace the cap before ejection (Figure 10.2). Rectification means that all the capsules are positioned cap uppermost in the machine ready for separation.

Several different methods have been developed for putting powder into the capsule body, but before this is examined in detail it is important to understand the problems that are likely to arise during the basic process. Firstly, a damaged capsule may block the feed system. Rectification may not take place, again resulting in a blocked feed system. Separation may not occur, so that the powder or other dosage

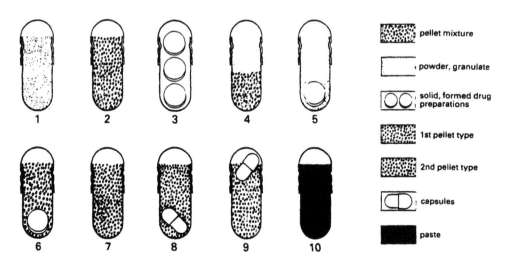

Figure 10.1 Various combinations of different drug formulations

176

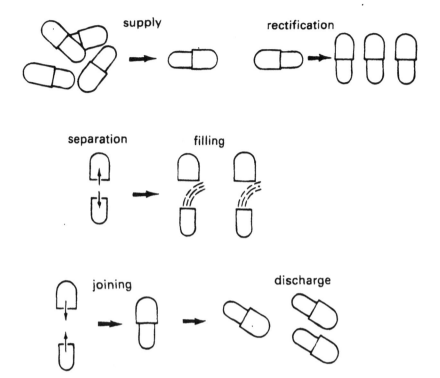

Figure 10.2 Capsule handling

form is available but cannot get into the capsule body. Transfer or selection of the dosage form may fail to take place. The powder may be wrongly delivered, or the alternative dosage form may be damaged during its entry into the capsule body. The cap or the body, or both, may be damaged when reunited. The body may be damaged on ejection from the machine. Finally, empty capsules may be ejected from the machine along with the filled ones.

All fully automatic equipment is fitted with devices that will eliminate some of these problems as the capsules pass through the machine. If new equipment is being evaluated, an examination of the above possibilities must be included in the test programme.

Capsule filling suffers from the disadvantage (compared with tablet making) that the method of filling the body of the capsule varies greatly from one manufacturer of filling equipment to another, and no one method has been universally adopted.

It is only in the last 25 years that machines producing in excess of 30 000 capsules per hour (cph) have become readily available, and only the last 15 that machines will fill over 100 000 cph. It is hardly surprising, therefore, that this dosage form has been generally ignored in favour of tablets.

Before examining in detail the filling methods currently employed by the leading equipment manufacturers, it is important to consider some of the powder characteristics that are required in a formulation for high-speed filling concomitant with the therapeutic requirements. The powder technology of the materials must define:

1. Bulk density and tapped density
   In a multi-component system it must show that for different batches of

177

chemically similar material the mixture is capable of being filled into the selected size of capsule.

2. Particle size

In general all formulations start as a mixture of fine particles, i.e. below 100 μm. A uniform particle size and particle density are required to optimize the mixing conditions before filling. Most drugs are milled to ensure that reproducible bio-availability and dissolution rates are obtained.

It has been demonstrated that particle size, crystal form, and method of formulation significantly affect the release of chloramphenicol (Aquiar *et al.*, 1968). Similar results were obtained with oxytetracycline (Brice and Hammer, 1969).

Flow

The formulation must flow on the filling machine such that filling rates of up to 3000 capsules per minute can be achieved. A typical formulation would contain the following:

Active ingredient
Filler, e.g. lactose
Flow aid, e.g. aerosil (Cab-O-Sil)
Lubricant, e.g. magnesium stearate
Surface active agent, e.g. sodium lauryl sulphate.

The object at this stage should be to produce a free flowing homogeneous powder with some cohesive properties. The requirement for cohesive properties will become clearer when the different filling methods are discussed.

It appears that little consideration is given to the properties that the powder must possess if it is going to be filled at high speed (above 25 000 capsules per hour) into capsules or removing damaged capsules before they enter the capsule handling system on a machine. For example, on a machine with a filling rate in excess of 2000 capsules per minute (over 120 000 per hour) damaged capsules at a level of 0.01% will interrupt the filling process or stop the machine every four minutes.

The powder technology used in designing a formulation capable of being filled at these speeds does not appear to have been studied at all. Equipment manufacturers have also neglected this aspect, and instead of designing mass flow hoppers have relied on using mechanical means of moving powders by fitting stirrers in the storage hoppers. Some workers, Irwin *et al.* (1970), Reier *et al.* (1968) and Ito *et al.* (1969), have studied problems in relation to specific machines, drugs, and compounds.

Irwin related the problems encountered during the development of a formulation for use on a Zanasi automatic capsule filling machine. He was able to show that a correlation existed between the flow properties when using a flowmeter developed by Gold *et al.* (1966) and the capsule fill weight variation. Reier *et al.* (1968) developed a mathematical model which related particulate properties, capsule size, and the operating rate of a semi-automatic filling machine to filled capsule character-istics. He also defined a number of parameters which are essential in order to ade-quately describe any process. Those which affect this filling process were the capsule size and the operating speed of the machine. Formulation variables include particle size distribution, specific volume, and flowability. Specific volume was obtained by placing approximately 50 ml of powder into a tared 100 ml graduate. This was tapped 15 times and the volume after tapping was divided by the weight of powder to give the specific volume in ml/g. Flowability was obtained by allowing powder to flow through a glass funnel, using a mechanical feeder, on to a calibrated vibrationless flat surface where it formed a cone. When the apex of the cone reached the funnel orifice, powder flow was stopped and the diameter of the base of the cone was determined. The area of the base was calculated rather than the angle of repose. This was termed flowability. This model is useful in its ability to provide information concerning the relationship of machine speed, capsule size, powder, specific volume, and flow to mean capsule fill weight from simple evaluations of small samples of the formulation to be filled. Ito *et al.* (1969) also used a semi-automatic machine, the Colton 8, to demonstrate that variation in machine conditions, formulation, and the type of auger used had considerable effect on the weight of fill and its coefficient of variation.

The angle of repose is unreliable when attempting to assess flow properties. This was demonstrated by Gold *et al.* (1966) who designed their own flowometer. This consisted of a strain gauge balance and recorder along with various hoppers. A vibrator was attached to one hopper to determine the rates of poorly flowing materials. Once the apparatus was calibrated the weight of material which flowed through the hoppers in a given time could be ascertained. This flowometer was used by Gold *et al.* (1966) to show the effect of glidants as defined by Munzel (1959). Gold *et al.* concluded that the flowometer gave a more reliable guide than angle of repose measurements to the flow rates of materials such as aspirin incorporating fused silicon dioxide, corn starch, and magnesium stearate as a glidant.

Gold *et al.* also found that the influence of glidants on spray dried lactose was negligible. The concentration is also important as quantities above 1.0% tended to decrease flow rates. Usually quantities in the region of 0.1% silicon dioxide were used.

According to King (1970) glidants are classified as a part of the lubricant class.

For filling on high-speed equipment it is essential to use a lubricant. Some materials like magnesium stearate may act as both glidant and a lubricant. One difficulty that arises is that there is no generally accepted method for measuring the lubricity qualities. For instance the USP and BP both give chemical requirements for magnesium stearate. It has been suggested (Butcher and Jones, 1972) that particle densities, packing characteristics, tensile strength measurements, and shear strength measurements may give a clearer indication of the lubrication properties of batches of magnesium stearate. Considerable batch to batch variation has been shown to produce large differences in compression properties (Hanssen *et al.*, 1970). Some materials such as corn starch commonly used as excipients in formulation work possess some lubricating properties. Others like lactose do not.

Instrumentation

The present author and his colleague (Cole and May, 1975) investigated some of these materials, using an instrumented Zanasi LZ/64 capsule filling machine. Strain gauges were fitted to a dosator system in an attempt to measure the compression forces which took place during the filling cycle. Two grades of lactose were investigated: microcrystalline cellulose (Avicel 101) and a modified corn starch (Sta-Rx 1500). Traces obtained when using these materials unlubricated and with 0.5% magnesium stearate were recorded (Figure 10.3).

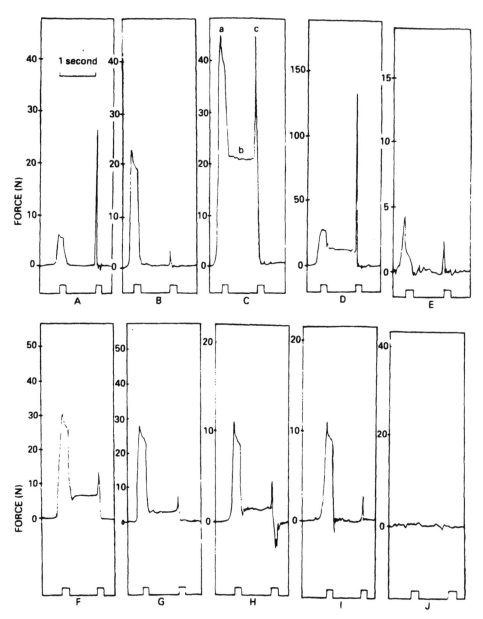

Figure 10.3   A. Lactose 80 mesh. B. Lactose 80 mesh + 0.5% magnesium stearate. C. Lactose 50T after 20 capsules. D. Lactose 50T after 50 capsules. E. Lactose 50T + 0.5% magnesium stearate. F. Avicel. G. Avicel + 0.5% magnesium stearate. H. Sta-Rx 1500. I. Sta-Rx 1500 + 0.5% magnesium stearate. J. Machine noise

With the lubricated materials, the machine was run for several revolutions before making recordings. With the unlubricated materials the hopper blades were run for several revolutions, but readings were taken immediately on fitting the dosator nozzles, to study the onset of binding of the dosator. Three main regions can be distinguished on the oscillograph traces:

180

- A force of 5.0–50 N representing compression of powder into the dosator.
- A retention force during carry over of the plug to the capsule body.
- An ejection force as the plug is ejected into the capsule. This can vary from 1.0 N to more than 350 N for an unlubricated material.

With unlubricated materials the ejection force from successive operations can be seen to increase rapidly, and it is sometimes accompanied by a concomitant fall in compression force. This must be due to some conditioning of the inner surface of the nozzle under certain conditions.

Various features of the force-time curves have not, as yet, been explained; for example, the general difference between individual materials and the fall of the force trace below the zero level just after ejection, a phenomenon most marked with unlubricated Sta-Rx 1500.

Filling Methods

There are a number of methods which are employed on capsule filling machines to ensure a uniform dosage in each capsule. No single method produces identical results, nor has universal acceptance. The latest methods used by the major manufacturers are very similar and depend on the principle of compressing the powder to form a plug or pellet. This is then ejected into the capsule body.

These machines have their origins in the nineteenth and early twentieth centuries when simple machines were constructed for filling a number of capsules in one unit operation. Today there are several variations of this method, the simplest comprising a flat plate which may be split into two halves; the upper half retaining the caps and the bottom half the bodies of the capsule shells.

Figure 10.4 illustrates a block containing three capsule bodies, a quantity of material, in this case pellets, filling them, and an excess covering the top of the block. If all the material is to be filled into the bodies then the tamping device will have to be used. According to the density of the powder several tamping actions may be required. The powder will have been pre-weighed to give a required weight of fill and spread evenly over the block, although

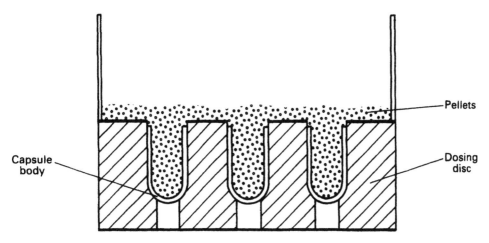

Figure 10.4 Plate method

when the powder has been formulated for a certain size of capsule and is of reasonably uniform density it is possible to use the plate method without pre-weighing. In this case excess material is spread over the capsule bodies until they are full and the excess removed before replacing the caps. Some form of vibration on the block is useful, despite the possibility of inducing segregation. Good formulation will minimize this. Various machines have been developed that use this principle, such as the Tevopharm and the Bonapace.

A development of the plate method was the use of a hopper fitted with a stirrer and an auger. An example is shown in Figure 10.5. The stirrer is used to feed material into the auger which in turn feeds it into the capsule body.

Hofliger and Karg produced a machine using this type of feed system which filled at the rate of 50 per minute. It operated only while the hopper was positioned above the capsule body, and the timing control on the auger drive regulated the fill weight. The hopper was fixed and the capsule bodies were moved by the plate.

In the Colton 8 machine the hopper was moved over a rotating ring containing the capsule bodies, and this enables a skilled operator to fill up to 20 000 capsules/hour.

Hofliger and Karg have developed a method, illustrated in Figure 10.6, based on the use of free flow of material into the body combined with a tamping piston. Powder initially flows into the holes in a dosage disc, which is machined to provide a certain fill weight in the capsule. A tamping punch compresses the powder against the base plate and then rises. Filling of more powder and retamping takes place in five successive stages. At the end of the fifth stage the dosage hole moves off the base plate and the plug of powder is ejected into the capsule body.

Difficulties can occur when using this method of filling if the batch-to-batch variation in density of the powder is very wide. The dose is controlled primarily by the thickness of the transfer disc, the adjustment of the tamping plungers and the depth of powder in the dosage hopper.

The compression method shown in Figure 10.7 is used by Zanasi (Zanasi is now part of the IMA Group of Bologna, Italy) and by mG2. The

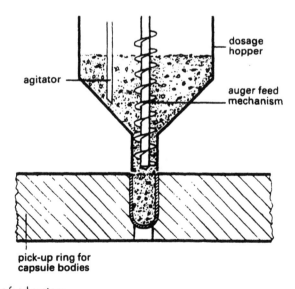

Figure 10.5 Auger feed system

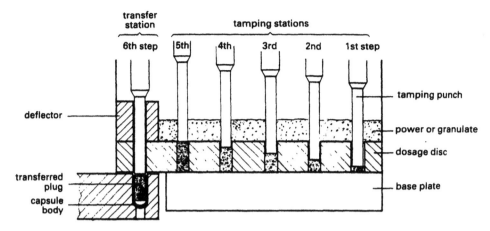

Figure 10.6 Tamping method

illustration is the dosator of a Zanasi LZ/64 intermittent machine. Powder is fed into the dosage hopper and its level adjusted to above twice the depth of the compressed plug. The dosator tube enters the powder bed and the powder inside it is pressed by the dosage piston just sufficiently to form a coherent plug that can be lifted by the dosator, carried to the capsule body, and pushed out into it by the piston. If the powder has been correctly formulated the use of the compression head is not necessary. The compression force used should be just sufficient to allow clean transfer to the capsule body and to ensure that the plug does not break up on

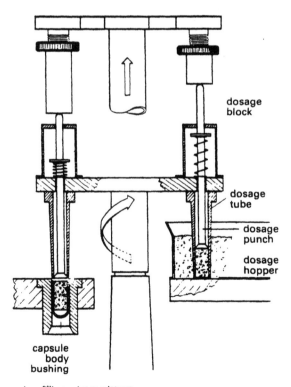

Figure 10.7 Compression filling – intermittent

Pharmaceutical Production Facilities

ejection from the dosing tube. The plug should also fit into the capsule body without protruding too far above the top. The dosator assembly is shown in more detail in Figure 10.8.

The calibration scale is used to set up each dosator assembly to the identical point at the commencement of a filling operation. On larger machines there are many dosators: on the model AZ-60, for instance, there are 24.

The fill weight is adjusted by regulating the height of the dosing piston inside the dosing tube. The depth of powder in the dosage hopper will also affect the fill weight. Extra compression may be achieved by using the compression head (Figure 10.7) which is adjusted to come into contact with the top of the piston when the dosator assembly is at the bottom of its stroke and forming the plug.

To minimize fill-weight variation a lubricant such as magnesium stearate ensures that the plug does not bind on ejection from the dosing tube or stick to the end of the piston. A uniform density is necessary in the dosage hopper, and a flow aid such

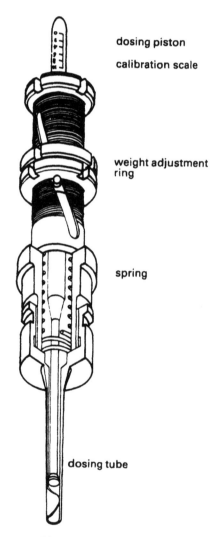

Figure 10.8   Zanasi dosator assembly

184

as aerosil (fused silicon dioxide) helps in this respect. Lubricants are often hydro-phobic, which is one reason for using minimal compression. The harder the compact, the more difficult it is to wet, and the more likely that the bioavailability of the product will be reduced.

The use of compression filling on a continuous machine was first developed by mG2 (Figure 10.9). Essentially there is little difference between the application of the principle on a continuous machine and on an intermittent machine. The dosage hopper here is annular, rotates, and is fed from a bulk hopper. It is possible to obtain much higher filling speeds using much simpler mechanical design providing that the turret carrying the dosators can be arranged so that there is no relative movement between it and the powder during the short time that the dosator is being dipped into the powder. For example, the Zanasi BZ-150 (claimed output 150 000 capsules/hour) has only one weight control and one pressure control, whereas the AZ-60 (claimed output 60 000 cph) has 24 dosators, which will all need individual adjustment. A comparison of the mechanical drive of the two machines shows similar advantages for the continuous action machine.

A European patent (Application No. 79302044.7) was applied for by Drugpack Ltd, a British company, for a different method of filling capsules. The essential fea-tures are shown in Figure 10.10. This machine was not developed commercially. However a number of its features were used by other equipment manufacturers.

The method employed uses a vacuum, 11, to remove air through an exhaust passage, 3, from the capsule body, 6. This causes powder, 5, to flow from the hopper, 4, down the inclined delivery tube, 2, into the capsule. The flow of material ceases after restoration of normal atmospheric pressure. The vacuum system is adjusted to cause the predetermined quantity of material to be dispensed into the capsule.

The points at which the passages 2 and 3 open in the underside of plate 1 are sufficiently close together to come within the mouth of capsules larger than size 5. The underside of plate 1 has a filling process. A pin, 10, raises the capsule body from a carrier, 8, against this sealing ring and activates the filling process. A filling machine has been built on this principle with an output of up to 20 000 capsules per hour.

There are a number of advantages using this technique. It is possible to fill single substances, which reduces the amount of formulation work required. It eliminates the use of powder lubricants which would ensure an improvement in bioavailability.

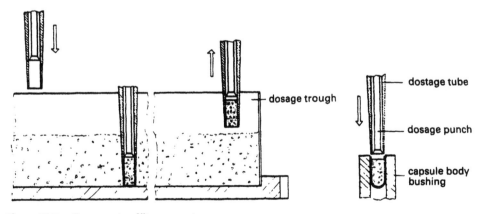

Figure 10.9 Compression filling – continuous

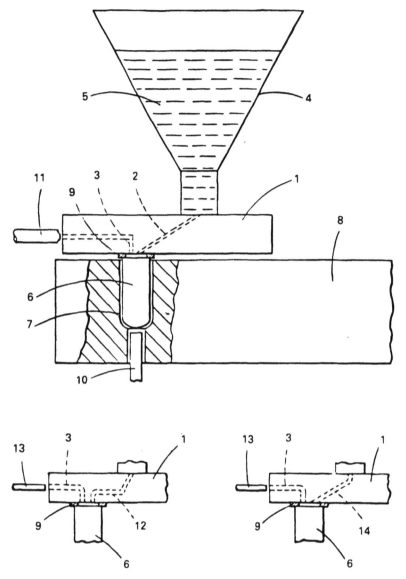

Figure 10.10   Drugpack principle (reproduced from Patent Application)

In the last 10 years a large number of products designed to provide prolonged therapeutic action have been introduced. They have been variously described as 'prolonged action', 'sustained release', or 'controlled release', which all imply an extended period of action for a given drug. One way of producing this type of action is to make coated pellets (time pills or time beads).

Provided that products act in the intended manner there are considerable therapeutic advantages to be obtained. Lipowski (1938) in British and Australian patents describes a dosage form where a number of small beads containing the dose of drug, with several different thicknesses of coating, are used, conferring the considerable benefit of a slow but constant rate of supply of drug to the patient. A similar dosage form using time release pellets has been patented by Blythe (1954). It enables

186

undesirable side effects to be controlled. The prevention of missed doses because of a patient's non-compliance is another advantage. Sustained release drugs are, therefore, a significant therapeutic advance in dosage design, and when used in a clear capsule can present an attractive product image. There are also some undesirable features inherent in a sustained-action product. Although a pellet may be coated so as to release part of the drug in the stomach and the remainder in the intestine 3 to 4 hours later, the period the pellet remains in the stomach can be anything from 30 minutes to 7 hours (Nelson, 1963). Some drugs are also less effective in divided doses, for example, *p*-aminosalicylic acid in the treatment of tuberculosis.

In general, when evaluating a pellet filling device the most important points to consider are the dosage accuracy (Marquardt and Clement, 1970) the extent to which pellets are damaged at various filling rates, the degree of segregation, if a mixture is used, the effect of batch variations in pellets on attainable filling speeds, the type of flow from the storage hopper, which ideally should be mass flow, the effect of ambient conditions on the filling characteristics, and the filling speeds at which partial fills start to occur.

As the speed of filling increases, the dwell time (i.e. the time taken for the dosing tube to dip into the powder and compress the plug of material) becomes shorter. This means that the powder must be of uniform density, easily compressible and not too elastic. The formulation must be free-flowing, but yet possess some cohesiveness.

It can be seen that the speed of the machine, the level of fill, and the density of the substance to be filled all have an influence on the dosage accuracy. One factor which considerably affects the fill-weight variation is the behaviour of the powder left in the hopper after a plug of material has been removed. Where the powder is free-flowing and non-cohesive the cavity will collapse and a simple stirring device will produce a homogeneous mixture before the next plug is removed. In cases where the material is very cohesive the powder must be thoroughly mixed before the next entry of the dosator.

Perry Industries, Hicksville, USA, developed a method based on its idea of using vacuum to dose vials with volumetric quantities of powders. The machine is continuous in operation and similar in action to the mG2 models. The principle, illustrated in Figure 10.11 is to draw the powder into the dosator by suction, applied through a filter pad.

Some compression does take place, but not to the same extent as in the Zanasi piston dosator. The material is held in place by the vacuum until the dosing tube is in position over the capsule body, when the powder is ejected by releasing the vacuum and applying positive pressure. This method has interesting possibilities because it does not rely on the movement of mechanical parts during the filling operation. Lubricants are not needed in the formulation. It is also possible to fill single substances, especially where the drug has a high dose level, although it is claimed that doses as low a 10 mg can be dispensed. This was the original machine which incorporated a 'no capsule no dose' feature. This was activated when capsules were not separated or when an empty capsule feed-tube became blocked and prevented a capsule entering the holding bush. The powder which had been taken into the dosing tube was blown out by a blast of compressed air, activated by an electronic sensor and relay, into the dosage trough, preventing loss of material and ensuring a clean machine.

The dosing tubes consist of two parts, a cylinder and an adjustable piston, made of polyethylene fitted with a nylon filter. Adjusting the piston alters the volume of

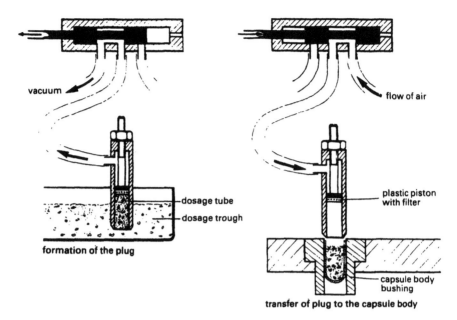

Figure 10.11 Vacuum filling method

powder that is picked up. The assembly is connected to the vacuum and air pressure system by a snap-closure connector.

Each of the individual dosage tubes has to be independently calibrated, central adjustment not being possible. The instrumentation allows a sample to be taken from any individual dosing tube without stopping the machine. The Zanasi, mG2, and Perry machines all depend on presenting a powder with a uniform constant density to the dosator. All three use an annular trough which is fed from a larger hopper, so it is important to know what the behaviour of the powder will be under the conditions that exist in these three machines.

One of the latest units for capsule filling to be introduced is the mG2 model G37/N manufactured by the mG2 company of Bologna, Italy. This is a rotary continuous motion machine with an output in excess of 100 000 cph. It provides attachments for the automatic filling of powders, pellets, and tablets or a combination of these types of dosage form.

The machine is suitable for use with a centralized 'in-house' vacuum and compressed air system. It is equipped with a fully interlocked hardened glass protective screen, and sound-proofed to 80 dBA.

The working cycle is shown in Figure 10.12. During the capsule rectification, open capsules, capsules with two caps, and capsules which fail to orientate are automatically ejected without interruption of the filling cycle.

Capsules are retained in a slot in the transfer disc by vacuum applied through a hole in the slot. This reduces the possibility of crushed capsules during the transfer to the filling hopper. Powder is maintained at a constant level in the central filling hopper by the high and low level probes fixed in the wall of the hopper. Unopened capsules are automatically rejected.

The annular powder filling hopper is designed such that the dosator picks up powder from a fresh position in the powder bed on each cycle. This enables the

Capsule-filling Machinery

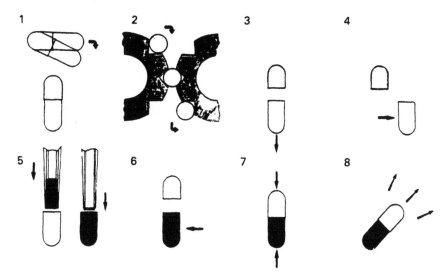

1. Capsule rectification (orientation)
2. Transfer
3. Separation
4. Alignment for filling
5. Filling
6. Alignment for closing
7. Closing
8. Ejection

Figure 10.12   Working cycle of mG2 Model G37N capsule filler

density of the powder bed to be maintained homogeneously and reduces the weight variation of the filled capsules. The NT/1 computer station enables samples of capsules from individual dosators to be selected and weighed.

The model G37 capsule filler may be used in conjunction with several pieces of equipment which improve the overall performance of the capsule filler such as the empty capsule pre-sorter model SV/LV1.

This unit is equipped with an automatic loader to the capsule filler. It is schematically shown in the layout plan, Figure 10.13.

The model SC filled capsule selecting machine enables the filled capsules to be inspected when ejected from the capsule filler, dedusted, and any empty and broken capsules removed.

The VR/1 computer statistically controls the capsule weight during the working cycle of the capsule filler. It is programmable and individual weights from each dosator are included in the printout. These weights are summarized as required by the programme, and the number of samples weighed are quoted together with the average weight, coefficient of variation, and standard deviation. Each sub-lot is recorded and the total for the day summarized. A feedback to the machine can also be provided to control the weight and the pressure exerted to form the plug during the filling cycle.

The importance of this unit relates to the hard copy produced showing the weight of individual capsules and the mean weight of random selected samples. This

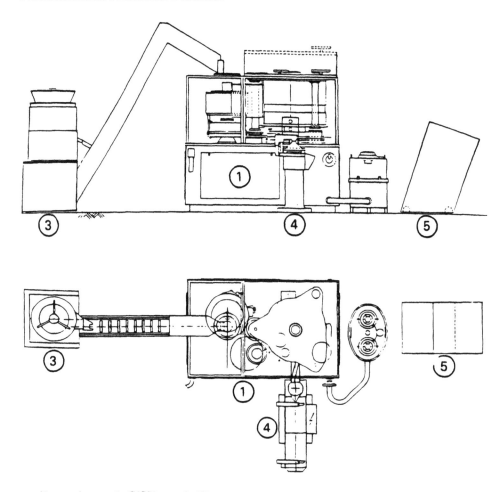

Nomenclature: 1. G37/N capsule filler
2. NT/1 Dosator sampling device*
3. SV/LV1 Empty capsule pre sorter
4. SC Filled capsule sorter
5. VR/1 Computer incorporates NT/1
*Not shown but is used to connect computer to filling head on machine.

Figure 10.13 Schematic of G37 capsule filler and ancillary equipment

printout and the summary of the day's production can be used to attach to analytical batch records. The balance used to weigh samples is a conventional analytical balance modified to provide an output to a desk type computer.

An alarm system and automatic stopping of the capsule filler may be fitted which alerts the operator whenever a capsule weight is outside the tolerance limits. The capsule filler is automatically stopped when the production quality is outside the pre-set limits for weight variation.

The frequency of the record print-out is variable and is included in the programme when the system is started. Such frequency will vary according to the operating conditions and the limits required by the control programme.

A machine with a similar operating cycle is the Zanasi BZ/150 manufactured by F. Zanasi SpA of Bologna, Italy. Three of these machines have been incorporated into the semi-automatic plant at the Cramlington plant of Merck Sharp and Dohme in the United Kingdom. This plant is controlled from a mimic board, Figure 10.14, and has a capacity in excess of 3.6 million capsules per day. A flow diagram is shown in Figure 10.15. The retail value of this product in 1990 was in excess of £150 000 per day. This plant operates in a strictly controlled temperature and humidity environment.

One of the characteristics of gelatin is its variable hygroscopicity, an important point with regard to the production of capsule preparations. The three following types of effect have to be considered:

- Effect of humidity on the capsule wall.

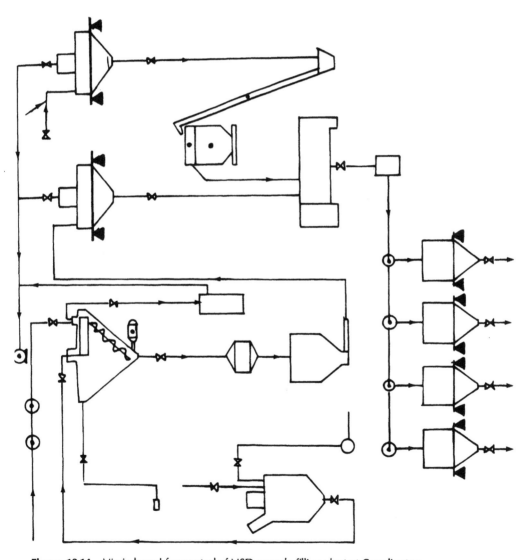

Figure 10.14 Mimic board for control of MSD capsule filling plant at Cramlington

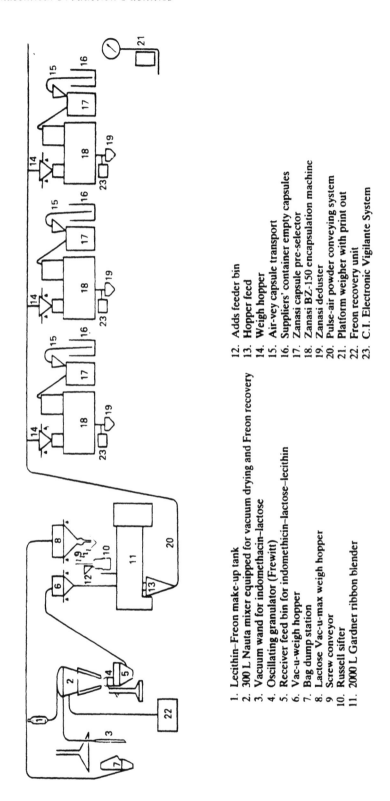

1. Lecithin–Freon make-up tank
2. 300 L Nauta mixer equipped for vacuum drying and Freon recovery
3. Vacuum wand for indomethacin–lactose
4. Oscillating granulator (Frewitt)
5. Receiver feed bin for indomethicin–lactose–lecithin
6. Vac-u-weigh hopper
7. Bag dump station
8. Lactose Vac-u-max weigh hopper
9. Screw conveyor
10. Russell sifter
11. 2000 L Gardner ribbon blender
12. Adds feeder bin
13. Hopper feed
14. Weigh hopper
15. Air-vey capsule transport
16. Suppliers' container empty capsules
17. Zanasi capsule pre-selector
18. Zanasi BZ-150 encapsulation machine
19. Zanasi deduster
20. Pulse-air powder conveying system
21. Platform weigher with print out
22. Freon recovery unit
23. C.I. Electronic Vigilante System

Figure 10.15 Flow diagram, MSD Cramlington

- Effect of humidity on the capsule contents.
- Effect of moisture contents on the capsule wall.

If gelatin capsules are stored in rooms where the relative humidity is 35–70% and the temperature 15–30°C, they will eventually reach an equilibrium water content of 10–16%. Under these circumstances, the elasticity and hardness of the capsule wall remains virtually unchanged. If the relative humidity were to fall below 30% then the capsule wall would lose an appreciable part of its water content and the capsule as a result would become so hard and brittle that it would be difficult to process mechanically. By contrast, relative humidity values in excess of 70% cause gelatin capsules to soften; here, too, mechanical processing would be impossible. It is claimed that if, however, these capsules are spread out in thin layers and dried, they will eventually again become suitable for filling, provided that they have not deformed as a result of sticking together.

These findings have been confirmed by Kuhn (1963) and Leupin (1964). Working independently, both authors have shown that storage under extremely dry conditions has remarkably little effect on the weight of gelatin capsules. Leupin found, for example, that 77 days' storage at 25°C and 20% RH led to a weight loss of only 2% whilst exposure to an RH of 60% was followed by a 1% increase in weight.

To minimize the effect of relative humidity on the elasticity of the capsule wall, gelatin capsules are generally packed in airtight, impregnated paper sacks and transported in sealed containers, which take the form either of drums of reinforced paper or of corrugated-cardboard boxes.

SOFT GELATIN CAPSULES

The History

A soft gelatin capsule is essentially a one-piece elastic container primarily designed to deliver unit doses of oily liquids. They have been available since 1833. Originally they were made in leather moulds in an elongated shape with a drawn-out tip at one end. This permitted cutting off the tip for injection of liquid inside the hollow gelatin shell. The opening would then be re-sealed with a small amount of molten gelatin.

Even in the early days a plasticizer had to be added to the gelatin to provide flexibility to the shell. Glycerin is still the substance of choice for this purpose, accounting for about 25% of the weight of the elastic shell.

Not only bad tasting medicine appeared in soft gelatin capsules in the early days. Easily insertable vaginal and rectal suppositories of vitamin products in the thirties did more to encourage the popularity of soft gelatin capsules than any other factor. This demand had to be matched by an innovative technical development. An engineer in Detroit, R. P. Scherer, devised the rotary die process, that made possible high volume production at the same time as the film is formed. The process is a little more complicated than that for producing the simple hard gelatin capsules, but the machine is neither so big nor so cumbersome.

Scherer's invention, as with Colton's hard capsule machine development, revolutionized the market for soft gelatin capsules. Relating his development to the vitamin market, it is interesting to note that his company also became a leading producer of bulk vitamins in the forties.

Pharmaceutical Production Facilities

MANUFACTURING PROCESS

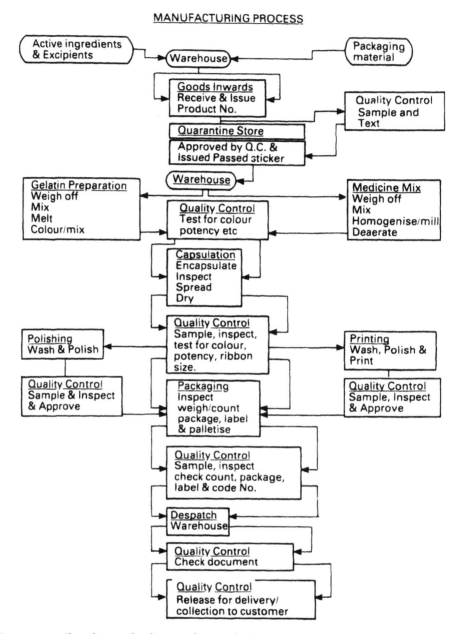

Figure 10.16 Flow diagram for the manufacture of soft gelatin capsules

The Process

Having the supply of material to be filled readily at hand is, of course, essential to a soft gelatin capsule manufacturer. Scherer's rotary die process comprises the following:

● Two gelatin ribbons are prepared, automatically and continuously, and fed simultaneously with the liquid or paste medicament to the encapsulating mechanism.

194

- The capsules are simultaneously and continuously filled, formed (the force of the injection of the medicament between the two gelatin ribbons causes the gelatin to expand into the die pockets to form the shape of the capsule), hermetically sealed, and automatically 'cut out' between two rotary dies. The resulting gelatin netting is collected for re-cycle.

- The resulting capsules are automatically conveyed to and through a naphtha wash unit and an infrared drying unit which partly dries the capsule. The partly dried capsules are spread on trays, and drying is completed in the tunnels immediately behind the equipment.

Figure 10.16 is a flow diagram which illustrates the steps in this manufacturing process.

Ideally, for maximum production efficiency, the continuous processes demand 24 hours per day, 5 days per week continuous operation. Thus formulations must be so designed as to maintain their desired physical characteristics, during this period as well as during periods of weekend shutdowns. The production capacity of each of these machines is determined by:

1. Die size, which determines the number of die pockets on the standard-size die plate, rotary die, or reciprocating die
2. The speed of the machine
3. The physical characteristics of the material to be encapsulated.

The capsule shell is composed of gelatin, a plasticizer, and water, and may contain additional or optional ingredients such as preservatives, colouring, and opacifying flavourings, sugars, acids, enteric agents, and medicaments to achieve desired effects. All materials and resulting products must be in conformance with the Food and Drug Laws.

The gelatin used is USP grade with additional specifications required by the capsule manufacturer. The additional specifications are concerned with the Bloom strength, viscosity, and iron content of the gelatin used.

References

AQUIAR, A. J., WHEELER, L. M., FUSARY, S. and ZEHNER, J. E., *J. Pharm. Sci.* **57**, 1844–50 (1968).

BRICE, G. W. and HAMMER, F. H., *J. Am. Med. Ass.* **208**, 1189–90 (1969).

IRWIN, G. M., DODSON, G. J. and RAVIN, L. J., *J. Pharm. Sci.* **59**, 547–50 (1970).

REIER, G., COHN, R., ROCK, S. and WAGENBLAST, F., *J. Pharm. Sci.* **57**, 660–6 (1968).

ITO, K., HITOMI, M., KAGA, S. J. and TAKEGA, S., *Chem. Pharm. Bull. Tokyo* **17**, 1138–45 (1969).

GOLD, G., DUVALL, R. N. and PALERMO, B. T., *J. Pharm. Sci.* **55**, 1133–6 (1966).

GOLD, G., DUVALL, R. N., PALMERO, B. T. and SLATER, J. G., *J. Pharm. Sci.* **55**, 1291–5 (1966).

MUNZEL, K., *First Industrial Pharmaceutics Conf., Land of Lakes, Wisconsin*, (1959).

KING, R. E., *Remington Pharmaceutical Sciences* 14E., p. 1652, Mack Publishing Co. (1970).

BUTCHER, A. E. and JONES, T. M., British Pharmaceutical Conference, September 11–15 (1972).

HANSSEN, D., FUHRER, C. and SCHAFER, B., *Pharm. Ind., Berl.* **32**, 97–101 (1970).

COLE, G. C. and MAY, G., *J. Pharm. Pharmac.* **27**, 353–8 (1975).

LIPOWSKI, I., *Austrian Patent 109,438, British Patent 523,594* (1938).

BLYTHE, *US Patent 2,738,303, British Patent 765,086* (1954).

NELSON, E., *Clinical Pharmacology Therapy* **4**, 283 (1963).

MARQUARDT, H. G. and CLEMENT, H., *Drugs Made in Germany*, Vol. 1, 21 (1970).

KUHN, T., *Pharm. Ztg. Berl.* **108**, 130 (1963).

LEUPIN, K., *Pharm. Ind. Berl.* **26**, 524 (1964).

11

The Design of a Sterile and Aseptic Manufacturing Facility

As she said this, she came suddenly upon an open place, with a little house in it about four feet high.

Manufacture and control of sterile medicinal products. Sterile products must be manufactured with special care and attention to detail, with the object of eliminating microbial and particulate contamination. Much depends on the skill, training, and attitudes of the personnel employed. Even more than with other types of medicinal products, it is not sufficient that the finished product passes the specified tests; and in-process quality assurance assumes a singular importance.

The Orange Guide.

Sterile products may be classified into three categories according to their manner of production:

- Those which are sterilized in their final containers (terminally sterilized). This is the method of choice whenever possible.
- Those which are prepared under aseptic conditions from previously sterilized materials.
- The system where sterility for manufacturing is of prime consideration but the product being manufactured is highly potent and can cause adverse reactions in operators.

Terminally sterilized products should be manufactured in a 'clean area'. This is a room with defined environmental control of particulate and microbial contamination constructed and used so as to reduce the introduction, generation, and retention of contaminants within the area.

The object should be to prepare a pyrogen free product with low microbial and particulate counts. The efficiency of a sterilization process depends on keeping the microbial load of the product to a minimum.

Non-terminally sterilized products, i.e. those prepared under aseptic conditions from previously sterilized materials, can be processed in clean areas until they have been sterilized but must subsequently be filled into the final sterile containers in 'aseptic areas'. An 'aseptic area' is a room, suite of rooms or, more usually, a special area within a 'clean area' designed, constructed, and with process utilities engineered and used with the intention of preventing microbial contamination of the product.

Table 11.1 Basic environmental standards for the manufacture of sterile products

Grade	Final filter efficiency (as determined by BS 3928)[1]	Recommended minimum air changes per hour	Max permitted number of particles per cubic metre equal to or above:[2] 0.5 μm	Max permitted number of particles organisms per 0.5 μm	Max. permitted No. of viable organisms per cubic metre[2,3]	Nearest equivalent standard classification		
						BS 5295[4]	US Fed. Std. 209B[5]	VDI 2083, P.1[6]
1/A (Unidirectional air flow work station)	99.997%	Flow of 0.3 m/s (vertical) or 0.45 m/s (horizontal)	3000	0	Less than 1	1	100	–
1/B	99.995%	20	3000	0	5	1	100	3
2	99.95%	20	300 000	2000	100	2	10 000	5
3	95.0%	20	3 500 000	20 000	500	3	100 000	6

Air pressure should always be highest in the area of greatest risk to product. The air pressure differentials between rooms of successively higher to lower risk should be at least 1.5 mm (0.06 inch) water gauge.

[1] BS 3928: Method for Sodium Flame Test for Air Filters, British Standards Institution, London, 1969.

[2] This condition should be achieved throughout the room when unmanned, and recovered within a short 'clean up' period after personnel have left. The condition should be maintained in the zone immediately surrounding the product whenever the product is exposed. (Note: It is accepted that it may not always be possible to demonstrate conformity with *particulate* standards at the point of fill, with filling in progress, due to generation of particles or droplets from the product itself).

[3] Mean values obtained by air sampling methods.

[4] BS 5295: *Environmental Cleanliness in Enclosed Spaces*, British Standards Institution, London, 1976.

[5] US Federal Standard 209B, 1973.

[6] Verein Deutscher Ingenieure 2083, P.1.

Such a room should comply, in the unmanned state, with conditions specified for Grade 1/B. (Table 11.1). With people present and work in progress Grade 1/A (Table 11.1) conditions should be maintained under contained work stations where products are exposed and aseptic manipulations carried out.

The basic standards required for aseptic and clean areas are given in BS 5295/ 1976 entitled *Environmental Cleanliness in Enclosed Spaces* and in the *Guide to Good Pharmaceutical Practice* 1985. They are summarized in Table 11.1.

Air Pressures

In the environment where sterility is the only consideration the air pressure is generally an outflow from the inner rooms through airlocks to atmosphere. The differential pressures between the rooms are maintained at 1.0 mm water gauge or greater.

When considering the handling of a toxic substance in the sterile area, a new set of problems exists. The outside air must not flow into the area, as in the sterile requirement. However, the inside area with potential toxic contamination cannot flow to the atmosphere. This is accomplished by placing the airlock at the highest pressure, thereby allowing air to flow to the atmosphere and into the inner rooms and then to the atmosphere through the HEPA filtration system.

The sterile air supply to the rooms is a constant volume supply, and the pressure is regulated by the position of dampers in the exhaust systems for the rooms.

It is necessary to know the status of all openings such as doors and pass throughs into the rooms. Any one of these being open for an extended period can effect the potential sterility of the room and containment of toxic materials.

Methods of Sterilization

The official methods of sterilization are described in the *British Pharmacopoeia* 1980, Addendum 1983, Appendix XVIIIA p A 56. to which reference should be made together with the *Pharmaceutical Codex* 1979, p 849. The methods described by the *British Pharmacopoeia* are:

- Heating in an autoclave
- Heating with a bactericide
- Dry heat
- Filtration
- Exposure to ethylene oxide·
- Exposure to ionizing radiation.

However, note that BP Addendum 1986, page iv, says:

> Reference to the process of heating with a bactericide has now been removed from all specific monographs of the *British Pharmacopoeia*. The section on this process will therefore be omitted from the Appendix on Methods of Sterilization in the next full edition of the *Pharmacopoeia*.

General Requirements for the Design of Aseptic Areas

Sterile medicinal products must be prepared in specially designed and constructed manufacturing departments which are separate from other manufacturing areas, and

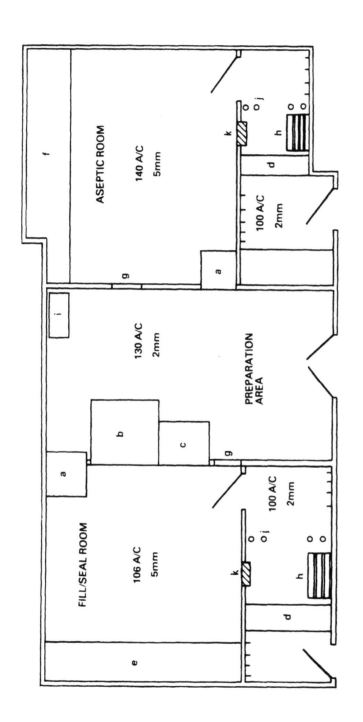

a = transfer hatch
b = autoclave
c = oven
d = step-over barrier

e = vertical laminar flow bench
f = horizontal laminar flow bench
g = viewing/speech panel
h = gown racks

i = water still
j = vertical chromium barrier tubes
k = wash basin/dryer

A/C = approximate air changes per hour

mm = air pressure in mm water (above atmospheric pressure)

Figure 11.1 Sterile products unit

in which the different types of operation such as component preparation, solution preparation, filling, and sterilization are effectively segregated one from another. A floor plan is shown in Figure 11.1.

The general requirements are outlined in this chapter. Reference should be made to the *Guide to Good Pharmaceutical Manufacturing Practice* 1983, BS 5295: 1976, *Rules and Guidance for Pharmaceutical Manufacturers* (MCA) and *Aseptic Pharmaceutical Manufacturing* Books I & II.

The Sterile Suite:

1. The surfaces of walls, floors and ceilings should be smooth, impervious, and unbroken in order to minimize the shedding or accumulation of particulate matter, and to permit the repeated application of cleaning agent and disinfectants where used. Bare wood must be avoided.

2. To reduce accumulation of dust and to facilitate cleaning, there should be no uncleanable recesses and a minimum of projecting ledges, shelves, cupboards, equipment, fixtures and fittings. Coving should be used where walls meet floors and ceilings in aseptic areas (and preferably in other clean areas as well).

3. Pipes and ducts should be installed so that they do not create recesses which are difficult to clean. They should be sealed into walls through which they pass. Ideally they should be kept out of the area and only protruding through the wall at the point of use or into a specifically designed panel.

4. False ceilings must be adequately sealed to prevent contamination from the space above them.

5. Drains should be avoided whenever possible. If installed they should be fitted with an effective, easily cleanable trap and an air break to prevent back flow.
   Sinks should be of stainless steel, without overflow, and be supplied with water of at least potable quality. Drains and sinks should be excluded from aseptic areas.

6. Access to aseptic areas should be restricted to authorized persons who enter through airlocked changing rooms (Figure 11.2) where normal clothing is exchanged for special protective garments. These rooms should be flushed with

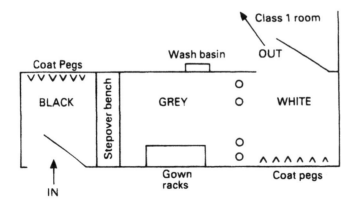

Figure 11.2   Changing room

filtered air at a positive pressure lower than that in the clean or aseptic areas. They should be used for personnel only and not for the passage of materials or equipment.

Hatchways and airlocks used for the passage of materials and equipment into clean and aseptic areas should be interlocked so that only one side may be opened at any one time. Sliding doors should be avoided since they are difficult to clean.

The Environmental – Air Supply

To fulfil the requirements of Class 1 or Class 2 conditions the areas have to be supplied and continuously flushed with air of suitable quality and at positive pressure. It is usually introduced into the room through HEPA filters of appropriate efficiency.

In Figure 11.1 the air supply to the two Class 1 rooms enters through the back (horizontal flow) at a pressure of 5 mm water gauge above atmosphere or roof (vertical flow) of the laminar flow work stations at a pressure of 5 mm water gauge and leaves via vents to the Class 2 room and to the changing rooms (pressure 2 mm water gauge) flowing in from the inner door and venting to the outside via the entrance door. This ensures that, on entry, and when following the changing procedure (see below) personnel move against the air flow and into progressively cleaner conditions (from black to grey to white). For toxic material handling the pressurization will be designed to ensure that no escape of the substance is permitted through the airlock.

The air supply to the Class 2 room enters via vents from the Class 1 rooms and is removed through two ducts in the ceiling. It is extracted at such a rate that the pressure in this room is 2 mm water gauge above atmospheric.

The extracted air is recirculated together with some outside air which is then passed through an air conditioning unit where humidity and temperature are controlled.

The plan of the facility shown in Figure 11.1 indicates the air pressures, number of air changes per hour, and direction of air flow.

Equipment

1. Equipment should be designed and installed so that it may be easily cleaned, disinfected or sterilized as required.
2. Items of equipment should be kept to the minimum required for a particular process.
3. Equipment and materials liable to generate particles or fibres must not be taken in to clean aseptic areas.
4. Articles required for aseptic areas should be sterilized and passed into the area through double-ended sterilizers or airlock hatchways.

Personnel

1. High standards of personal hygiene and cleanliness are essential. Any conditions which may cause the shedding of abnormal numbers and types of micro-organisms should be reported, e.g. colds, skin infections.
2. Only the minimum number of personnel should be present when work is in progress, and their activities restricted to avoid excessive shedding of micro-organisms and particles.
3. Personnel should receive adequate training in the disciplines relevant to the manufacture of sterile products.

Clothing

1. Operators and their garments are, potentially, the most significant sources of microbial and particulate contamination.
2. Personnel entering clean or aseptic areas must change into special garments which include head and foot wear. These garments should shed virtually no fibres or particulate matter and retain particles shed by the body. They should be comfortable to wear and loose fitting to avoid abrasion.
3. In aseptic areas personnel should wear sterilized single or two-piece trouser suits, gathered at the wrists and ankles with high necks. Headgear of helmet/cowl type must totally enclose the hair and beard and be tucked into the neck of the suit. Footwear should totally enclose the feet and trouser bottoms should be tucked inside. Powder-free rubber or plastic gloves should be worn with the garment sleeves tucked inside them. A non-linting face mask should also be worn.

Changing Procedure

1. Access to aseptic areas should be through rooms where normal clothing is exchanged for the protective garments described above.
2. Changing and washing should follow a written procedure (standard operating procedure).
3. Changing rooms (Figure 11.2) are usually divided into a
   (i) black area – where normal protective garments (e.g. lab coats) are removed, separated by a step-over bench from
   (ii) grey area where hands are scrubbed and protective garments donned
   (iii) white area where overboots are put on and, if necessary, sterile gloves.

Cleaning Procedure

Both the *Guide to Good Pharmaceutical Manufacturing Practice* and BS 5295: 1976 indicate that clean, aseptic, and other related processing areas should be cleaned frequently and thoroughly in accordance with a written programme. There must be a complete plant housekeeping and management system.

Arrangements should be made for:

- Routine daily cleaning before work commences.
- Periodic cleaning of the walls, fittings, and working surfaces, etc., and after cleaning the room should be tested for acceptability before full working is resumed.
- The control and cleaning of protective garments including periodic checks to ensure that the cleaning procedures do not produce particulate contamination products.

Where disinfectants are used, different types should be employed in rotation to discourage the development of resistant strains of microorganisms. Disinfectants and detergents should be monitored for microbial contamination.

Monitoring

BS 5295: 1976 states that provision should be made for monitoring the following environmental conditions:

- Particulate contamination
- Temperature and humidity
- Air velocity and pressure.

The *Guide to Good Pharmaceutical Manufacturing Practice* gives limits for the maximum number of viable organisms per cubic metre for each class of room, and the monitoring of microbial contamination will have to be carried out.

The Guide states: 'Areas should be frequently monitored microbiologically by means of "settle" plates, surface sampling, air sampling, or other appropriate methods. The monitoring should be performed whilst normal production operations are in progress. Records should be retained and immediate remedial action taken as soon as results deviate significantly from those usually found in the area concerned.'

a. Changing procedure for Class 1 rooms
   Black area

   1. Enter the first changing area, wearing clean plimsolls.
   2. Remove outer clothing and jewellery but keep shoes on.
   3. Sit on stepover bench and put on blue plastic overshoes as you swing each leg over. Do not allow overshoes to come into contact with floor in black area.

   Grey area

   4. Wash hands and arms up to elbows for 1–2 mins at least, paying particular attention to washing between the fingers and under the nails. The wash-basin controls should be foot operated.
   5. Dry hands under the automatic hand dryer.
   6. Put on hood and face mask. Check that no hair is showing. Bearded men should wear a visor hood.

7. Select an appropriate size suit. These are two-piece suits and should be put on as follows:

- Select the top half of the suit and put it on by pushing the sleeves through with the hands, then pulling it over the head. Fasten it up ensuring that the hood is tucked under it at the neck.
- Step out of the shoes (but leave the overshoes covering them).
- Put on the bottom half of the suit by pushing the legs through with the feet, one at a time without letting the garment touch the floor. Tuck the top half flap inside the bottom half and fasten the appropriate studs.
- Seal wrists and ankles with the press stud fastenings.
- Step back into the shoes (still covered by the overshoes).

By following this procedure it is possible to put the suit on without letting it touch the floor.

White area

8. Step through the barrier (poles) putting on overboots, and keeping the covered shoes on.
9. Put on gloves and pull these over the sleeves of the suit.
10. Spray the gloved hands with 70% alcohol.
11. Pass into the aseptic area.

To leave the room the reverse procedure is adopted. The protective clothing and boots should be brought out of the changing room and put in the laundry basket.

b. Changing procedure for Class 2 room
Before entry into the Class 2 room an operator must:

1. Be wearing a clean white overall.
2. Be wearing plimsolls.
3. Put on a cover cap ensuring that no hair is showing.
4. Put on a pair of disposable PVC overshoes.

Each size of garment and overboots can be coloured according to size.

Size Colour Code for Garments and Overboots

The suits are two-piece, and a size larger than the chest size should be worn if an intermediate size is required.

Suits	34″	38″	42″	44″
	White	Midblue	Apple green	Navy
Hoods	Small	Medium	Large	
	White	Midblue	Apple green	

Most females will require 'small' and most males 'medium' sizes.

Visor hoods – for bearded males – Navy (one size only).

Overboots

Shoe size	4/5	6/7	8/9	10/11	12/13
	White	Midblue	Apple green	Navy	Primrose

Cleaning Procedure

The daily cleaning and periodic cleaning of the rooms detailed in the standard operating procedure has to be carried out before the area can be sterilized for manufacturing operation.

Containers for Sterile Products

This section is concerned only with the packaging of sterile medicinal products rather than sterile medical devices, such as dressings, pacemakers, etc., in other words products of a pharmaceutical nature covered by the Medicines Act 1968. However, mention will be made of sterile empty containers sold for subsequent aseptic filling, e.g. 3LTPNA bags. The purpose of sterile packaging is to contain, identify, and protect and deliver a dosage form in sterile conditions to a patient. The requirements for containers for sterile products can be summarized as:

1. They are chemically compatible with the product.
2. They withstand sterilization.
3. They maintain sterility of the product.
4. They permit safe withdrawal of the product.

The most important requirement of a container is to contain the product safely and to protect sterility of the product. The product may be a liquid, a solid, or a gas. The *European Pharmacopoeia* defines a container as:

> A container of pharmaceutical use is an article which contains or is intended to contain a product and is, or may be in direct contact with it. The closure is part of the container.

The container may be a single dose container, or it may be a multi-dose container. A single dose container holds a quantity of the preparation intended for total or partial use as a single administration. A multi-dose container holds a quantity of the preparation suitable for two or more doses. These containers can range in the case of large volume parenterals from rigid and flexible to glass and plastic bags or bottles, to low volume parenterals where ampoules, vials, prefilled syringes, or other novel devices are used. Where irrigation is required, then generally glass or plastic in the form of sachets is used. For eye drops glass and plastic bottles and droppers are used. Eye ointments require some form of tube generally manufactured from a plastic material. Historically, glass containers have been the most important, but plastics have seen a tremendous increase in use and popularity in recent years.

The EP highlights three classes of glass for the manufacture of containers for injectable preparations, and these classes are based on the hydrolytic resistance of the glass.

Class Type	Description	Use
I	Neutral glass. Borosilicate glass. High hydrolytic resistance due to its composition. Used to make blown containers and in the form of tubing to make vials. Expensive.	SVPs regardless of pH. Mildly alkaline LVPs or where thermal shock resistance is important.
II	Sulphated soda glass. High hydrolytic resistance due to surface treatment. Dates back to 1930s. Moderate cost.	LVPs. Some SVPs – usually neutral or acidic. Not intended for re-use.
III	Glass having moderate hydrolytic resistance. Soft glass.	Not intended for aqueous preparations. Used for some oily injections and dry powders for reconstitution.

Hydrolytic resistance is defined as the resistance offered by the glass to the release of soluble mineral substances into water under prescribed conditions of contact between the interior surface of the container and freshly distilled water.

Some of the advantages of glass are:

* Good chemical resistance (depends on the glass type).
* Neither absorbs nor exudes organic ingredients.
* Impermeable. With proper closures entry or escape of gases (water vapour) is negligible.
* Easily cleaned.
* Transparent – facilitates inspection of contents.
* Rigid, strong and dimensionally stable.
* Resists puncture.
* Will hold a vacuum.
* Can be autoclaved at 121°C or may be sterilized by using dry heat sterilization.

Some of the disadvantages of glass are:

* Breakage during sterilization, particularly soda-lime glass.
* Attacked by alkaline solution.
* The development of hair-line cracks in transit – allowing ingress of moulds.
* Much heavier than plastics.
* Requires venting during administration.
* Vials require sealing by closures of a different material, giving rise to problems during autoclaving.

- Requires inspection and washing before use.

Plastics

With the development of polymer technology over the last 30 years, plastics have become alternatives to glass for LVP packaging. They have also been used on a limited scale for SVPs, but recent developments in polyolefins for SVP containers may increase their application in this area.

Some of the advantages of plastics are:

- Relatively unbreakable.
- Light (less than 1/10 the weight of glass).
- Easily fabricated.
- Cheap.
- Possible to eliminate the need for pretreatment.
- Single use.
- Small filling ports – less chance of contamination during filling.
- Possible to completely seal by fusion.
- Machines are available to blow-mould, fill and seal in a continuous operation.

However, the polymers used in parenteral containers must be carefully chosen to avoid the following disadvantages:

- Clarity is in most cases inferior to glass, although translucency and even transparency can be attained.
- Cannot match the barrier properties of glass to moisture and oxygen.
- Attention must be given to additives present in the plastic due to potential leachability into the medication.
- Sterilization of fluids in plastic packs poses problems since the container must be protected from deformation and bursting.

Each polymer has its own characteristics and limitations. However, by the use of additives it is possible to improve a particular polymer's performance.

The polymer's formulation must not include any substances that can be extracted by the contained product so as to alter the stability or efficacy of that product or increase its toxicity.

The general classes of additives most commonly found in plastics used for LVP packaging are:

Antioxidants
Heat stabilizers
Lubricants
Plasticizers
Fillers
Colourants

They can be in liquid, solid, or fine particle form and concentrations vary between 0.01% and 60% w/w.

Antioxidants. The oxidative effects of heat, light, ozone, and mechanical stress in the presence of oxygen cause the formation of free radicals, and these contribute to the degradation of the polymer with the loss of physical and mechanical properties. The presence of antioxidants reduces the degree of degradation and extends the life of the container.

Examples: phosphites and thioesters

Heat Stabilizers. During the manufacture of PVC, degradation products will catalyse degradation reaction or cause discolouration. This can be reduced by the addition of heat stabilizers.

Examples: metallic stearates and epoxidized plasticizers

Lubricants. Used to modify the surface characteristics and aid processing.

Examples: fatty acids, silicones

Plasticizers

Used to improve workability, flexibility, impact strength, and resilience. However, they reduce tensile strength.

Examples: dialkyl phthalates

Fillers. Used to improve flexibility, impact resistance, dimensional heat stability, and to reduce cost. In parenteral PVC containers small amounts of submicrometre fillers are used as brighteners and co-colourants without impairing transparency.

Colourants. Used to mask aging colour changes. Ultramarine blue is one of the most commonly used.

Owing to the possible complexity of some of the formulations in use the potential for problems of toxicity and interaction can be appreciated.

The EP states that the nature and amounts of the additives are determined by:

Polymer type
Manufacturing process
Use of the article.

Antistatic agents and mould release agents are forbidden in the EP.

Acceptable additives are indicated in the EP specification given for each material.

Some examples of commonly used materials, their properties, and manufacture are indicated below:

Polyvinylchloride, PVC. PVC is the major polymer resin in the vinyl family, and although it is the most widely used polymer in the medical industry, its medical use accounts for only a very small fraction of the world's annual production.

PVC first became important in the 1940s as a synthetic replacement for rubber.
The medical uses of flexible PVC include:

IV tubing
Drip chambers
Catheters
Blood bags

LVP containers

PVC was the first polymeric material to replace glass for blood and LVP containers. It is produced by polymerizing vinyl chloride gas in the presence of an initiator.

Residual vinyl chloride (VC) monomer can cause toxicity problems, and in 1975 the FDA proposed regulations to restrict the use of VC in contact with food products. Since then the PVC industry has expended great effort to keep vinyl chloride monomer levels to less than 1 ppm.

There is no single standard formulation of medical PVC. Each is designed for a specific purpose, and many are proprietary; however, some general guidelines are discussed here.

Component	Concentration (parts per hundred parts of resin by weight)
PVC	100
Plasticizer	30 to 40
Stabilizers	0.25 to 7

Only a very few of the commercially available stabilizers are suitable for medical applications, these include the calcium–zinc types. Organo-tin compounds are not liked in the UK.

Epoxides, such as epoxidized soya bean oil, are commonly used as co-stabilizers. *Lubricants are important.* Plasticizers modify the normally rigid PVC into a flexible plastic, but plasticizer leachability is of considerable concern, and there is current controversy surrounding the use of

di-2-ethylhexylphthalate
(DEHP or DOP)

At present no consensus has been reached about whether DEHP is a mutagenic hazard to humans, and it continues to be used, but some alternatives are being investigated and some manufacturers appear to be moving away from PVC altogether.

Flexible PVC bags are fabricated from either separate webs or lay-flat tubing by high frequency welding, and it is a relatively simple process to seal into the bag administration and additive ports. Production of the containers should take place in an appropriate environment, such as a clean room.

Since PVC film is relatively clear when not hydrated; inspection of the fluid within is possible. It is also possible to print directly onto PVC for labelling purposes.

The moisture transmission rate of PVC is higher than that of polyolefins and consequently the filled bag must be overwrapped before sterilization. Under normal storage conditions up to 2% water loss per month could occur from an unprotected bag, thereby concentrating the contents. For this reason the overwrap must be left on until just before use.

Polyethylene (Polythene)

Polythene is a suitable material for the fabrication of parenteral containers provided that relatively high density material is used. This has a sufficiently high melting temperature to withstand sterilization but suffers from a loss of pliability and an increase in opacity.

Containers can be readily made by a blow-moulding process to produce a sealed unit which is particle free and ready for filling and resealing. However, it is difficult to fabricate 'no touch' ports into these containers, and since it is not practicable to print onto polythene, labelling must take place post sterilization, eliminating the possibility of overwrapping before sterilization to produce a sterile outer surface on the container.

It is also necessary to protect the container in a rigid sleeve to prevent deformation. This can affect heat transfer.

Loss of moisture is much less of a problem than with PVC.

Since some degree of rigidity is inherent the drip rate may fall as withdrawal of liquid takes place, leading to improvised airway access being made with concomitant risk of contamination.

Few additives are used (and compatible) with polythene, and a list of permitted stabilizers is given in the EP.

Polythene has definite advantages in the packaging of alkaline infusions such as sodium bicarbonate 8.4% w/w since PVC is permeable to CO leading to deterioration of the product, and some glass types are susceptible to alkaline attack.

Closures: Elastomeric Components

Elastomers or 'rubbers' can be moulded into an almost limitless variety of shapes to meet specific needs, and the desirable properties are compressibility and resealability respectively.

Properly formulated elastomeric closures seal small voids in mating surfaces such as inside vial necks, and they are easily penetrated by hypodermic needles to reseal after withdrawal.

A full and detailed paper on elastomeric closures for parenteral products by Smith & Nash appears in *Pharmaceutical Dosage Forms: Parenteral Medications* Vol II.

As with PVC, this is a complex formula and again the requirement is that these materials do not include in their composition any substance that can be extracted by the contents in such quantities as to alter the efficacy or stability of the product or increase its toxicity.

Some of the types of ingredients used in pharmaceutical rubbers are:

• Elastomer or polymer
• Curing/vulcanizing agent
• Accelerator
• Activator
• Antioxidant
• Plasticizer
• Filler
• Pigment.

No one formulation can meet all pharmaceutical needs, and the relationship between component supplier and dosage form manufacturer is very important.

The following factors affect the selection of a rubber closure:

- The drug
- The vehicle
- Preservative if any
- pH of the product
- Buffer system
- Metallic sensitivities
- Moisture/gas protection required
- Configuration of the closure
- Colour.

It may be the case that improvement in one property, e.g. reduction in curing, causes losses in another, e.g. worse gas permeability.

Labelling

Identification is usually achieved by some form of labelling, and the purpose is to ensure that medicinal products are correctly described and are readily identifiable.

Medicines Act Leaflet 42 (MAL 42) is a *Guide to the Medicines (Labelling) Regulations 1976* (SI 1976 No. 1726) and deals with the labelling of containers and packages of medicinal products. In addition MAL 49 *Notes on the Medicines (Labelling) Amendments Regulations 1977* (SI 1977 No. 996) adds requirements for appropriate warnings and other information to be displayed on some medicines.

MAL 42 part 3 gives the standard particulars which include:

- Name of product
- Pharmaceutical form
- Quantitive particulars.

These requirements are eased for small containers (less than 10 ml), and requirements for ampoules are the same as for other small containers except that the route of administration must always be stated.

Reference should be made to the relevant MALs and regulations for full details.

MAL 42 does not really cover in-process labelling, and it is necessary to consider that aspect further.

The product needs to be identified both during manufacture, as in intermediate produce, and for release for sale, as a finished product.

Appendix 5 of the *Guide to Good Pharmaceutical Manufacturing Practice* 1983 (*The Orange Guide*) begins with the statement:

> Mislabelling (or the mis-use, mis-application or mix-up of other printed materials) is the most frequent cause of product hazard and product recall. Constant care and attention should be given to preventing such errors at all stages.

Measures to help avoid mislabelling are then given. This applies to sterile and non-sterile products but is particularly important in the case of parenteral products.

A particularly serious incident in 1985 re-emphasized the need for careful control and security during the labelling, filling, and inspection of parenteral products.

The incident resulted in a batch of freeze-dried injectable product being withdrawn, and seven points were highlighted in the *Medicines Act Information Letter* No. 48 of October 1986 as being worth including in the design of control systems. The article was titled 'Safety measures in the manufacture and control of filled but unlabelled parenteral products'. The seven points can be summarized as:

1. Fill into labelled containers, or label immediately after filling. If this is not possible temporary labelling measures are strongly recommended. Each container should be identified by a method which differentiates it from products in similar containers or of similar appearance. Suitable methods include:

   - Direct printing on containers
   - Printed secondary labels
   - Numerical codes
   - Ring codes
   - Bar codes.

2. If colour is used in the control system, test operators for colour blindness.

3. Process only one size of one product in an area at any one time.

4. Seal and label transit containers of unlabelled product with:

   - Product code
   - Product name
   - Strength
   - Lot number.

5. Eliminate mislabelling risk and reconcile quantities of filled, unlabelled product at each stage by sensitive procedures. Do not release product unless discrepancies are satisfactorily investigated.

6. Do not return in-process samples to the batch unless by documented and sensitive reconciliation procedures supervised by both production and QC.

7. Reinforce with training programmes.

Labelling Methods for Finished Products

Labelling can include all printed material accompanying the product, including the carton, package insert, and the actual label affixed to the immediate container.

Common methods of labelling immediate containers:

Direct:	Screen or litho	– PVC bags and ampoules
Indirect:	Ceramic printing	– glass
	Paper labels	– polythene containers
	Vinyl labels	– ampoules

Once labelled some containers become difficult to inspect, and some container materials are difficult to label, e.g. polythene. Some manufacturers' solutions to these problems are given here.

Label Style

Style is not addressed in the regulations except in the sense that:

'All particulars should be clear, legible and readily discernible'.

Label printing should be indelible.

Style is left to the manufacturer, but could be challenged by the licensing authority which has a policy discouraging the use of colour coding on patient user labels especially for supply into NHS hospitals.

Container Protection

The product requires two forms of protection:

- From the physical contaminants with the environment.
- From the chemical contaminants and degradation processes that can be induced by the environment.

The patient also requires a convenient and safe container for handling the dosage form.

The product needs protection from:

- chemical contamination
- particulate contamination
- microbiological contamination

and also

- physical protection during sterilization, subsequent handling, storage, and transit.

That the parenteral container must be completely sealed is well accepted. What is not agreed is to what extent this should or could be demonstrated during routine production testing. Most official compendia make no reference to leak testing, and until recently there was little published guidance.

A recent publication by the Parenteral Society addresses these issues in the case of glass ampoules: *The prevention and detection of leaks in glass ampoules*, Parenteral Society Technical Monograph No. 1.

The need is to make ampoules resistant to bacterial ingress, but chemical contamination and changes to the atmosphere within the ampoule are also of importance.

Perhaps it is useful to provide here a definition of a fault in an ampoule.

Any interruption to the physical structure of the ampoule that will permit the ingress or egress of materials or any undue stress or weakness that may develop into such a condition.

Leak testing must be embodied in the general concepts of GMP, i.e. that reliability cannot be 'tested into' a product, but is carried out in a consistent manner where it should indicate whether the failure rate is constant between batches and the examination of rejects can indicate possible sources of problems.

The production of leak resistant ampoules can therefore be summarized:

1. The causes of leaks in ampoules.

2. The prevention of leaks in ampoules.
3. Leak testing for overall QA.

Causes of leaks:

- Thermal cracks
- Mechanical cracks
- Faulty seals.

Prevention of leaks:

- Avoid glass to glass attrition
- Care in design of machinery
- Good control and design of the ceramic printing process
- Draw seal not tip seal
- Take care with filling and flame setting
- Take care with volatile contents
- Check seals for stress with polarized light
- Measure seal thickness to validate.

Tests for leaks – consider two in common use:

1. Dye intrusion challenge tests. A suitable test cycle would be:

   - Evacuate loaded chamber
   - Hold vacuum for dwell period
   - Admit dye solution under vacuum
   - Pressurize chamber to desired level
   - Hold pressure
   - Release pressure and remove dye
   - Wash ampoules
   - Inspect and sort (spectrophotometry and fluorescence).

2. High frequency spark testing: An automated method which detects leaks in defined areas of the ampoule (e.g. tip, base). The presence of a leak causes a change in a high frequency electrical signal placed across the ampoule. The method is restricted to aqueous products with an acceptably high conductivity.

Summary:

Method	Advantages	Disadvantages
Dye intrusion	Sensitive	Unsuitable for coloured products
	Tests whole ampoule	Contamination by dye
	Uses existing equipment	Subsequent clean
High frequency spark testing	Continuous – no contamination	Aqueous products only
		High capital costs
	Very sensitive – finds weak seals	Tests at set locations only

Depyrogenation

The most significant pyrogen for the pharmaceutical industry is endotoxins associated with the outer membrane of Gram-negative bacteria.

These toxins are constantly shed into the organism's environment. These compounds are heat stable and can survive ordinary steam sterilization cycles; however, various methods for the depyrogenation of packaging components are available, and the two most common are:

1. Dry heat treatment
   This is the method of choice for depyrogenation of glass and metal components and requires an exposure of not less than 250°C for not less than 30 min. The method of endotoxin inactivation is incineration.

2. Rinsing
   This is the oldest and simplest method of endotoxin removal from solid surfaces and uses rinsing with a non-pyrogenic solvent, usually sterile water for injection BP. Low levels of surface endotoxin contamination can be effectively removed from glass and stoppers with an appropriate wash-rinse procedure. Rinse water can be monitored throughout the process with the LAL test to validate endotoxin removal.

Protection from Particulate Contamination

Even if properly washed containers are filled with particle free solution under adequate environmental conditions, products in glass containers can become contaminated with particles. This may, of course, be due to the product formulation, but fluids in soft glass containers may contain flakes of glass.

Closures are the other main source of particulate contamination, such as:

- Fragments of rubber
- Carbon black
- Zinc oxide
- Chalk
- Clay.

All these may be released, as may mould spores.

Lacquering has been used (cured epoxy resins and more recently Teflon), but some lacquers may flake off into the product.

Protection from contamination during sterilization

A number of instances have been reported where bottle contamination was thought to have been caused by the introduction of contaminated spray cooling water in the autoclave chamber.

This can occur if the bottles are under partial vacuum at this point in the cycle time, and it is exacerbated by air ballasting in the chamber. The internal vacuum may draw in contaminated spray cooling water.

Also, contaminated water can backtrack under overseals. Design of the container/closure and sterilization cycle, and the use of sterile cooling water, is important.

Protection: Interaction with Rubber Closures

Rubber may react with drug formulations in the following ways:

Adsorption	– the concentration of a substance on the surface.
Absorption	– may follow adsorption. The substance passes through the surface and becomes distributed throughout the mass.
Permeation	– occurs when the above two stages are followed by transmission and loss through the closure.
Leaching	– is the reverse in that it is the phenomena of a substance migration from the closure into the product.

Lachman *et al.* (*J. Pharm. Sci.*, **51**, 224–232 (1962)) studied the adsorption of preservatives such as benzyl alcohol chlorbutol and the hydroxybenzoates into natural, neoprene, and butyl rubbers and found that natural rubbers exerted the least deleterious effect on the preservative content.

Wash and Chien (*PDA Report* No. 5, 1984) extended this by looking at partition coefficients of preservatives.

Lyophilization stoppers for freeze dried products are formulated from rubber with low vapour transmission. Usually butyl, or a halobutyl natural blend is used.

Delivery of Product

The EP states:

> The container is so designed that the contents may be removed in a manner appropriate to the intended use of the preparation.

Common routes of administration of sterile products have been covered earlier in the chapter.

The following is a discussion on some of the delivery systems for parenteral products. Some of these can be considered as an extension of the container system itself, and some are even an integral part of the container:

* Solution sets
* Burette sets
* Novel direct delivery systems.

Delivery of Intravenous (IV) Fluids

A paper in the *Pharmaceutical Journal* of 18 November 1985 by Furber *et al.*, titled 'Characteristics of containers for IV fluids' described a method for investigating the emptying characteristics of vented and unvented, rigid, semi-rigid, and flexible IV fluid containers under gravity. They concluded that the fully collapsible bag most closely compared to the ideal performance and had the advantage of being a closed system; however, it is difficult to establish how much fluid the patient has received, and some commercial containers contained considerable overages.

Regulatory Aspects and Published Standards and Specifications

In the UK those products which come under the Medicines Act 1968, i.e. full containers, and those which do not, for example, empty sterile containers such as 3L TPN bags sold for subsequent aseptic filling, must be distinguished between.

Finished sterile pharmaceutical products would normally require a product licence, and MLA 2 *Guidance Notes on Applications for Product Licences* (Rev. Dect 85) is relevant.

The application form MLA 201 requires details of the nature of the container and closure, and supporting data are also required.

Expert reports provided as part of the licensing process under EEC directive 75.319 require details of control of starting materials (including packaging) and stability in the chosen container system.

The report must give an opinion on how certain the identity of container materials is if they are tested to a specification developed along the lines of a pharmacopoeial monograph.

A spokesman of the licensing authority recently stated at an Industrial Pharmacist Group meeting that:

> Compatibility of the product with its container, and with diluents and their containers, has to be demonstrated. Plastic containers should be examined for sorption and leaching.

The licensing authority normally requires that all ampoules be leak tested and that where a dye bath test is not used the detection limits of the clearly defined alternative procedure must be known.

These are some of the requirements, but as with most licensing it is often a question of 'show us what you have done and we will tell you if it is acceptable'.

BS 5736 Pt 1 1979 *Evaluation of Medical Devices for Biological Hazards – Selection of Biological Methods of Test* provides good guidance for toxicologists working in this field, but additional requirements may be needed in order to comply with the Medicines Act.

The scope of this specification includes containers. The other points describe the test methods. Category E includes parenteral containers.

The following diagrams summarise the dangers, safeguards and interrelationships of the design and operation of a sterile manufacturing facility.

DANGERS

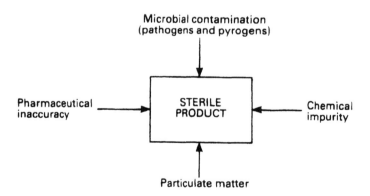

SAFEGUARDS

MICROBIAL

Room design, Air handling,
Filtration, People, Sterilization,
Raw Materials, Packaging,
Sterility testing, Record-
keeping, Batch-release data,
QA, QC, Validation

Pharmaceutical
Formulation, Stability,
Sterilization, Assay,
Raw materials, Packaging,
Record-keeping,
Batch-release data,
QA, QC, Validation

STERILE
PRODUCT

Chemical
Raw materials,
Assay, Packaging,
Manufacture,
Record-keeping,
Batch-release data,
QA, QC, Validation

PARTICULATE

Room design, Air-handling,
Filtration, Raw materials,
Packaging, People, Record-
keeping, Batch-release data,
QA, QC, Validation.

INTERRELATIONSHIPS

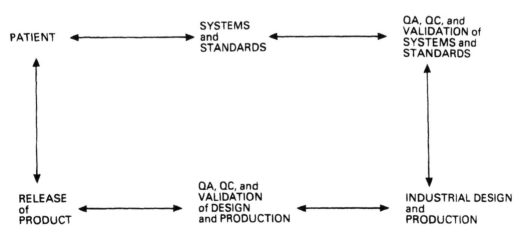

PATIENT

SYSTEMS
and
STANDARDS

QA, QC, and
VALIDATION of
SYSTEMS and
STANDARDS

RELEASE
of
PRODUCT

QA, QC, and
VALIDATION
of DESIGN
and PRODUCTION

INDUSTRIAL DESIGN
and
PRODUCTION

References

Specifications
  BP 80
  EP IV.2 Containers VI.2.2 Plastic Containers
  USP XXI Containers 661

DHSS Specification: TSS/S330.011 (Sept 81) *Specification for Pharmaceutical*
                        Bottles for Sterile Fluids
                        (Graduated and Non-Graduated Bottles)
                        TSS/S330.022 (Sept 81) Stoppers for above
                        TSS/S330.033 (Sept 81) Overseals for above

Leak Testing, etc.

The Prevention and Detection of Leaks in Glass Ampoules Parenteral Society Publication –
  Technical Monograph No. 1.
Sterility, Pyrogen, Particulate, and Package Integrity Testing (*Advances in Parenteral Sciences
  I*) Ed M. J. Akers, published by Marcel Dekker, Inc., NY, 1984.

Depyrogenation

Pyrogens: Endotoxins, LAL Testing and Depyrogenation (*Advances in Parenteral Sciences II*)
  FC Pearson, Marcel Dekker, Inc., NY, 1985.

Plastics (General)

Plastics Materials, fourth ed., Ed. J. A. Brydson, by Butterworth Scientific 1982.

Parenteral containers:

Microbiological Hazard of Infusion Therapy, Eds Phillips, Meers and D'Ary, MTP Press,
  1976.
Pharmaceutical Dosage Forms: Parenteral Medications Vol I and II, Eds. Avis, Lachman and
  Lieberman, Marcel Dekker, Inc, NY, Vol I 1984; Vol II 1986.

Regulatory aspects

Medicines Act Leaflets:
  MAL 2 *Notes on Applications for Product Licences.*
  MAL 18 *Licensing Requirements Involved in the Packing and Labelling of Medicinal
          Products.*
  MAL 42 *Notes on the Medicines (Labelling) Regulations 1976.*
  MAL 49 *Medicines (Labelling) Amendment Regulations 1977.*

Available from: Medicines Control Agency
                Department of Health
                Market Towers
                1 Nine Elms Lane
                London SW8 5NQ.

British Standards

BS 5736: Part 1: 1979   *Evaluation of Medical Devices for Biological Hazards.*
                          Part 1: Guide for the selection of biological methods of
                              test.

BS 795: 1983	British Standard *Specification for Ampoules.*
BS 5295: 1976	*Environmental Cleanliness in Enclosed Spaces,* British Standards Institution, London.
GMP Guides	*Guide to Good Pharmaceutical Manufacturing Practice 1983 (The Orange Guide).*

Guide to Good Manufacturing Practice for Sterile Medical Devices and Surgical Products 1981 (The Blue Guide).

Rules and Guidance for Pharmaceutical Manufacturers 1993 (MCA)

All available from HMSO.

Code of Federal Regulations Food & Drugs

Section 21 parts 200–299

also

Section 21 parts 600–799

Books

Aseptic Pharmaceutical manufacturing technology for the 1990s, Part I, Eds W. P. Olson and M. J. Groves; Part II, Eds M. J. Groves and Ram Murty, Interpharm Press.

12

Special Production Systems

'It's my own invention ...'

Tablets and their companion dosage form, hard-shell two-piece capsules, are the most widely used dosage form in the world. For the manufacturer and the patient the advantages of this type of product are considerable. A tablet is relatively simple to manufacture, e.g. aspirin tablets, and generally easy for the patient to take wherever he may be, in compliance with the medication regime. Of course, there are tablet and capsule products which are highly sophisticated, using such mechanisms as sustained release and the elementary osmotic pump principle, but even when so combined, these two dosage forms produce more than all the others put together.

Traditionally, the manufacturers of tablets and capsules have relied on many manual operations. Transfer of materials from one area to another and from one machine to another, has progressed very slowly. Material handling systems have been neglected in favour of human effort. Buildings were erected and equipment installed with little thought given to the links between each stage of the process and the way materials would flow through the facility. Times have changed even in the pharmaceutical industry, and now the cost of labour, the cost of the basic drugs, and the squeeze on profit margins have combined to ensure that facility design has had to adopt a more integrated approach. In the early 1970s Merck built two plants, one at Cramlington in England and a second at Elkton in the USA, using totally enclosed systems for material transfer and with a computer to control and monitor all the essential operations (Figure 12.1). These plants had design capacities of up to 1000 million coated tablets per annum for a single product. Similar plants were built on other European sites by Merck but with a multi-product manufacturing philosophy. These were operated on a campaign basis and required lengthy cleaning periods between production runs. All of Merck solid dosage products are relatively large in production terms, and the use of a worldwide integrated material management system enabled Merck to programme the campaign and achieve high levels of productivity. It enabled the company to dramatically increase capacity while maintaining, or even reducing, manning levels.

Pharmaceutical processes, like all manufacturing processes in these days of the microchip and programmable controller, are automated in order to enhance a number of operating parameters. Safety, reproducibility, productivity improvements, and most importantly quality are the principal benefits of a well designed automated facility. These processes span a variety of different pharmaceutical unit operations and have become standard in pharmaceutical production at

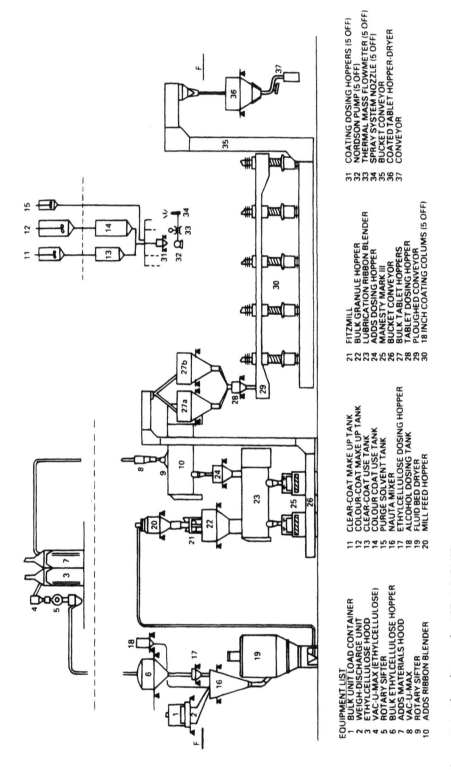

Figure 12.1 Flow diagram for FCT, ALDOMET

EQUIPMENT LIST
1 BULK UNIT LOAD CONTAINER
2 WEIGH-DISCHARGE UNIT
3 ETHYLCELLULOSE HOOD
4 VAC-U-MAX (ETHYLCELLULOSE)
5 ROTARY SIFTER
6 BULK ETHYLCELLULOSE HOPPER
7 ADDS MATERIALS HOOD
8 VAC-U-MAX
9 ROTARY SIFTER
10 ADDS RIBBON BLENDER

11 CLEAR-COAT MAKE UP TANK
12 COLOUR-COAT MAKE UP TANK
13 CLEAR-COAT USE TANK
14 COLOUR COAT USE TANK
15 PURGE SOLVENT TANK
16 NAUTA MIXER
17 ETHYLCELLULOSE DOSING HOPPER
18 ALCOHOL DOSING TANK
19 FLUID BED DRYER
20 MILL FEED HOPPER

21 FITZMILL
22 BULK GRANULE HOPPER
23 LUBRICATION RIBBON BLENDER
24 ADDS DOSING HOPPER
25 MANESTY MARK III
26 BUCKET CONVEYOR
27 BULK TABLET HOPPERS
28 TABLET DOSING HOPPER
29 PLOUGHED CONVEYOR
30 18 INCH COATING COLUMS (5 OFF)

31 COATING DOSING HOPPERS (5 OFF)
32 NORDSON PUMP (5 OFF)
33 THERMAL MASS FLOWMETER (5 OFF)
34 SPRAY SYSTEM NOZZLE (5 OFF)
35 BUCKET CONVEYOR
36 COATED TABLET HOPPER-DRYER
37 CONVEYOR

a number of facilities.

The systems replaced conventional granulating processes which involved manual dry mixing, granule formation in a planetary mixer, and tray drying operations. These conventional methods were labour intensive, time consuming, and produced dusty operations.

A typical Merck facility is located at Elkton, Virginia. Raw materials are loaded through receiving hoppers into product holding hoppers on the third floor of the manufacturing module. This loading is performed by operating personnel, and it is the only time operators are required to handle material throughout the manufacturing process. From this point, movement of materials is accomplished under computer control by either gravity or by pneumatic, belt or bucket conveying systems. In addition, all hoppers are on load cell systems to monitor and control the flow of materials through the process and to provide the information needed for mass balances and accountability.

From loading, the ingredients are discharged in the proper weight proportions into a ribbon blender on the floor below. After blending, the material is discharged to a granulating hold hopper on the first floor. The material is then fed to the enclosed granulator by a weight belt feeder. This piece of equipment continuously weighs the material on a given length of belt and adjusts the speed of that belt to ensure that material is fed to the granulator at a constant flow rate. The granulator itself is a liquids-solids 'zig-zag' blender. It was selected for this application because of its adaptability to the continuous flow process. The granulating solution is continuously sprayed through a rotating – intensifier – bar located in the blender. The zig-zag blender continuously rotates to provide for complete blending and to progress the material through the granulator.

Following this step, the granulated material is discharged onto a continuous throughput fluid bed dryer and then onto the lubrication, compressing, and the film-coating operation. Total throughput of granulation is about 5000 kg/16 hour day.

Several variables are important in producing a good granule, and they are therefore closely monitored and controlled. The control of some of these variables, such as solution pressure, intensifier bar speed, and the revolving speed of the blender have been determined and fixed. They are monitored primarily to ensure that no change has occurred. The major concern then is to ensure the proper control of liquid and solid material flow rates. Providing a constant flow of each will not necessarily produce a consistent granulation, since factors within each can widely affect the agglomeration process.

One way to measure this variation is through the amperage registered on the intensifier bar motor, since it varies directly with the force required to move the bar through the granulation. However, monitoring this parameter continuously only provides an early warning of change that has occurred.

To detect potential problems early is difficult. In this system, higher than expected material losses led to an analysis of the raw materials entering the process to determine what changes in them could most affect granulation. Surface area, as measured by tapped density and mesh analysis of the component, was identified as the most significant variable since it can vary by as much as 20% around the mean. Obviously, such significant changes can have a large impact on the ability of a fixed amount of granulating solution to agglomerate a constant weight of solid material.

Therefore, solution flow rate is now matched with both the surface areas of the solids, as well as their weight. The result of these changes was to reduce material loss in manufacturing by over 50%, since fines are no longer generated by inadequate solution. An important by-product of this approach was the elimination of all batches which required reprocessing. This process has achieved a high degree of reproducibility, enhanced product quality, and, by reducing losses, has demonstrated the benefits of an automated system.

At another of Merck's facilities where a second automated granulating system was introduced, the product was previously produced by conventional labour intensive methods.

Dust generation during manufacturing, and the large capacity required, led to the design of an enclosed automated granulating system. Here a batch process was developed which resulted in all dry mixing, granulating, wet mixing, and drying taking place in a single piece of equipment.

The automated system requires one operator to load raw materials into the loading hopper and to unload lubricated product into portable intermediate bulk containers (IBCs). It shares with another automated process two computer operators who are located in a central control room. In addition, manual back-up to the automated computer control was integrated into the plant design. This back-up was very helpful in the development phase and serves as a fail-safe in ongoing manufacturing. The granulating equipment chosen for this process was a 3000 L nauta mixer with dual vacuum and forced air drying capabilities. This type of equipment has now been replaced in today's 'state of the art' facility by the Spectrum (microwave dryer/mixer granulator) and the Topo granulator.

The facility was designed to provide 100% containment of product. Fork trucks or pallets never enter the facility; instead, the drums enter on roller conveyors and they are discharged through a rubbish chute after use (at the bottom of which is positioned a drum crusher and compactor). IBCs are dedicated to the process and remain within its walled boundaries. This total segregation of product eliminates the potential for product cross-contamination and may be especially useful when processing for toxic or other dangerous products.

Materials are manually loaded by an operator into the loading hopper, and conveyed pneumatically into the mixer. Under computer control the materials are granulated, and during this operation, the computer monitors the important processing parameters which include product weight, granulating solution addition, arm and auger speeds, vacuum pressure, jacket temperature, and product temperature.

On completion the granulation is pneumatically transferred through a comminutor mill for size classification into a blender. Lubricants are added and the resulting granulation dumped into stainless steel IBCs.

The computer control room that monitors the activities of the nauta mixer is used also to support other manufacturing activities.

Early in the 1960s Merck made the transition from sugar coating to film-coating. During that period, several new film-coated products were introduced. These early film-coating systems were scaled up from laboratory units. When larger facilities were needed for film-coating in the 1970s these conventional pans were supplemented by the purchase of additional more modern units.

At this time film-coating suspensions were manufactured manually with mills and mixing vessels driven by air agitators, and tablets were coated in conventional pans which rotated on rollers. Air was supplied directly onto the tablet bed and exhausted out of the back of the pan. Air atomized nozzles equipped with manual clean-out needles were used in each pan and fed with solution from a pressure pot. Pan speed, air pressure, supply air, and supply air temperature were controlled manually from gauges located on the front of the pan. Each batch of tablets coated weighed approximately 120 kg and had a cycle time of three to four hours. The tablets were then unloaded and transported to beeswax-lined copper pans for polishing and waxing.

To provide a 'state-of-the-art' facility an automated film-coating facility was installed, using pan coating.

Pans generally provide greater flexibility than column coating. The existing open tank suspension make-up system was discarded in order that an enclosed facility with direct piping to the new coating pans could be built. This concept of automated suspension delivery had already been proven at another facility and was adopted. An integrated clean-in-place system was included in the design, and a manual operating capability was integrated into the design.

At the core of this automated facility were six film-coating pans. The 'train' concept for film-coating solution delivery and product segregation was incorporated into the facility design, which allows the film-coating of three different products (one in each train) while the preparation of three additional suspensions is underway. Computer control provides for this 'reserve and request' situation only after satisfying certain conditions. They are:

- Process equipment is mated to the correct product number.
- Process equipment is not being used currently for another manufacturing process.
- Process equipment is in a clean state.

The automated aspects of this facility have replaced manual methods both for suspension preparation and also for the film-coating process itself. Built into the computer program are a number of critical parameters to the spraying process. Pan speed, suspension spray rate, air flow rate, exhaust air temperature, and pan pressure have been determined for each specific coating process. By a pre-selection of the proper mix of these film-coating parameters, an elegant, reproducible coated tablet is manufactured. In addition, if any critical parameter falls outside these pre-selected values, a controlled shutdown of the film-coating process takes place.

It should be emphasized that the shutdown is controlled, so that adverse elegance effects to the product being coated are minimized. After correction of the cause for the shutdown, spraying is then re-initiated.

As with the other automated processes, the control system for this facility is designed so that it may be operated either totally by computer, partly by computer and operator sequencing, or totally by operator sequencing. The control room operator is in charge of the system at all times, and his console serves as the communication between the computer and the field. This enables him to perform all functions of monitoring and communication necessary to 'remotely' operate the entire process.

A significant advancement in the approach to automated film-coating was made with the completion of this facility. Reproducible processing conditions; productivity improvements; enhanced tablet elegance and quality; improved materials handling and environmental conditions, were all achieved.

An additional MSD pharmaceutical manufacturing facility is located in Wilson,

North Carolina. Here the pharmaceutical product processing is performed in one of two manufacturing modules. Each module is equipped with a high shear granulator, a fluid bed dryer, in-line comminution for particle size reduction, and a ribbon blender, and supported by complete pneumatic conveyance systems through to the compressing machine.

The central mechanism for automated process control, however, is different from the previously installed systems. Process control at the Wilson plant site is afforded by a programmable logic controller (PLC). The PLC is a device which will perform logical decision making to achieve a desired task. One of the advantages of this system is its ease of programming. It is essentially a ladder diagram which controls relay circuitry. An engineer, electrician, or technician would find the logic of this programming easy to follow.

Thus, through the use of the PLC in concert with the process equipment, materials can be loaded, granulated, dried, comminuted, lubricated, and delivered to compressing by one operator. This operator's only manual task is to drop materials into a loading station and then press the appropriate buttons to complete processing.

Lastly, it is worth considering the alternatives, and there are examples of semi-automated processing which merit inclusion in a discussion on automation of pharmaceutical processing. It also demonstrates that one is able to achieve a great deal without the use of a microchip or programmable controller. Instead, timers and pneumatic switching devices to provide 'programmability' can be used.

One product uses a fluid bed granulating process, and the 'island of automation' concept illustrates how part of the system can be automated and integrated into the total plant.

This type of equipment requires large volumes of drying air and an exhaust system. Usually the intake air and exhaust air plenums are located on the roof of the manufacturing area, and this provides a quiet operating environment within the granulating room. Intake air is drawn into the granulating room where it travels across a steam coil for heating when required and then through a bag filter and HEPA filter.

The exhaust air fan draws air from the inlet air duct through the fluid bed granulator system. Before fluidization the inlet air duct is closed and a product loading valve opened. Materials (tablets or pellets can also be coated in this unit) are then charged into the fluid bed chamber, thus maintaining a completely enclosed processing environment.

Once loaded, materials are mixed, granulated, and dried in one unit operation. By selection of the appropriate granulating parameters (air temperature, degree of fluidization, spray rate, nozzle size, spray pressure, etc.), the desired particle size distribution can be achieved.

Additional particle classification may be needed, however, and in this particular process, approximately 5% of the batch was over or undersized. To maintain the dust-free environment within the granulating room, a pneumatic conveying system was installed which supplied granulation to an oscillating granulator for classification, from where it was transported to the blending area for lubrication.

This type of fluid bed granulator can be modified to provide for the column coating requirements of such dosage forms or intermediate products such as tablets, pellets, and granules.

Control of this system is operated under a semi-automatic mode where the oper-

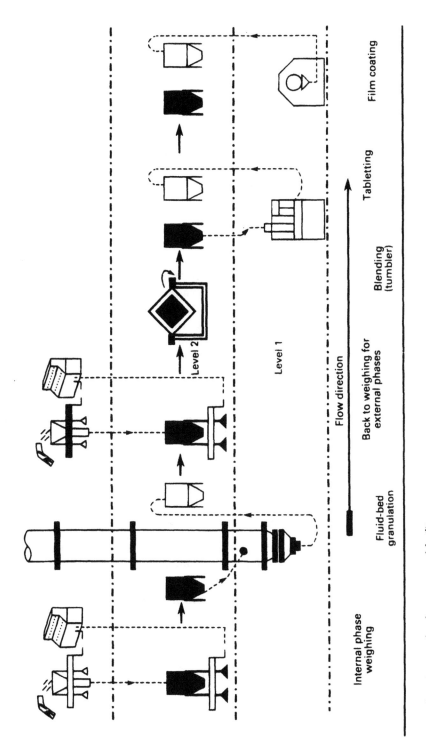

Figure 12.2 Flow diagram for the automated facility

ating panel is located within the granulating area. The operator is able to monitor and control the granulating or coating process by pre-selection of any number of operating parameters. No operator variables are introduced and a completely reproducible process is achievable.

An alternative approach was adopted in the early 1980s by Smith Kline & French (now Smith Kline Beecham) (Figure 12.2). Professor W. J. F. Lhoest and P. J. M. Froment designed and built a plant at Alcalá de Henares in Spain using as a basis concept the intermediate bulk container (IBC) or Flo-bin. The objective here was to use the IBC as a storage unit that could be moved from one stage of the process to the next using mechanical means, e.g.

- fork lift truck,
- automated guided vehicle (AGV),
- shuttle car,
- overhead monorail system,
- or hand operated mobile truck.

Each stage of the process could be activated as required and the material stored until a subsequent piece of equipment became available. This meant that a number of different products at different stages of manufacture could be processed simultaneously. It was claimed that many more products with different batch sizes could be produced, and only sections of the plant would require cleaning at any one time.

Subsequent to the SKF facility at Alcalá being completed, an improved version was erected in Milan, Italy, with an AGV transport system. Plants currently under construction or nearing completion with AGVs are those for Bristol Myers in the USA, Godecke in Germany, and Pharmachemie in the Netherlands. A number of other multi-nationals are considering similar ideas for plants in Germany, France, and Italy. These concepts will be described here.

General Design

Ideally, any production facility should meet the following objectives:

- Material whether in the form of powder, tablets, other dosage forms or packaging components, should move through the minimum distance possible before it is processed or manipulated in any way.
- Materials should be separated physically from the operative and the surroundings of the process areas.
- Losses of material should be reduced to a minimum.
- Be capable of 24 hour operation.
- Provide maximum flexibility to handle more than one product simultaneously, and a range of batch sizes depending on bulk densities.
- Low capital investment with low operating costs and easy expansion potential.
- Be capable of computer control, process monitoring, and batch recording.
- Comply with current good manufacturing practice.
- Easily validatable.

230

- Reduce analytical costs by standardizing and increasing batch sizes.

The IBC used in these plants is a large vessel designed to hold powders and engineered to allow the product granules or powder to flow out when the valve at the bottom is opened. Care must be taken with this design to ensure that a range of powders, granules, or dosage forms can be handled. Ideally, the IBC should be a mass flow container, but because of the cohesive nature of many of the materials used in pharmaceutical manufacture, some external means or internal flow activator may be required to ensure complete discharge of the container's contents. Alternative internal designs are necessary for handling tablets to minimize damage on impact when the dosage form enters an empty bin.

In the early stage of the process the IBC is used to mix various materials in a fine particulate form. This can be difficult as many materials have cohesive properties and have poor flow properties. It is useful to design the IBC to be able to mix batches of materials that occupy between 10 and 80% of the available volume. A considerable number of experiments will need to be conducted on the flow of single raw materials, mixtures, and granulations out of the IBC into the processing equipment to demonstrate its practicability. At this stage of the project a comprehensive powder characterization programme must be developed to optimize the powder technology of the materials at all stages of manufacture. Some parameters that should be measured are:

- Rate of flow from hoppers with varying sizes of discharge port, i.e. flowability.
- Angle of repose.
- Bulk density.
- Particle size range.
- The effect of batch-to-batch variation of raw materials on the flowability.
- The effect of storing batches of partly processed materials on discharge.
- The effect of using vibration to assist in discharging: i.e. does segregation result?
- Any effects of humidity, or other environmental parameters.

The Jenike Shear cell is a useful instrument for determining flow properties and the cohesive nature of raw materials and developing a flowability index.

To design a production system using the IBC requires a consideration of many aspects of pharmaceutical technology. Essentially it is a material handling problem and will require consideration of the following before selection can be made and a design commenced:

- automated guided vehicle systems (AGVs)
- pneumatic vacuum transfer systems
- complete IBC transfer systems
- discharge and receiving station systems
- automated container washing systems and clean in place systems (CIP)
- bulk and unit load conveyor systems
- blending
- batch formulation

- IBC weighing systems
- packaging
- automated warehousing
- loss in weight feeder systems.

By utilizing the selected systems, specific equipment and material gravity flow as far as possible, a flexible manufacturing facility can be achieved allowing easy reconfiguration for the manufacture of different products. With the addition of advanced computer systems and the automated supply of raw materials, a fully integrated computer integrated manufacturing (CIM) system can be installed, for both present and future expansion requirements. Automated warehousing can be coupled with the production materials handling systems to provide the complete solution, e.g. raw material supply and finished product storage, that is monitored and controlled from a central plant host computer system.

From the dispensary stage, where batches of materials are weighed out, and through to the actual packaging of the finished component, all materials are contained within closed systems. The materials handling system and associated equipment is an integral part of the process operation providing the medium for material flow in the correct sequential steps and subsequent handling at each process operation.

In the following sections a typical automated two-floor concept (as shown in Figure 12.2) oral dosage flexible manufacturing facility is considered and all the automated systems and equipment for materials handling are explained individually.

System Description

This system has a two-floor concept with a raised dispensary over the second floor level and a first floor technical area; the process rooms containing the manufacturing equipment are located on the ground floor.

The materials handling system is operated entirely at first floor level where product materials in batch lots are fed from the dispensary by gravity to the receiving station and into a product IBC. The charged IBC is transferred using an AGV transportation system to the various process stations in sequence on the first floor. IBC blending and IBC washing stations are located on this floor. IBCs positioned above the process room are discharged using gravity to process machines located in individual rooms on the ground floor.

Product IBCs are also filled on the first floor from the ground floor process machines using pneumatic vacuum transfer into an IBC located at a receiving station (RS) positioned adjacent to the corresponding discharge station (DS). Full IBCs are then collected and transported using an AGV to the next process step in the manufacturing sequence and then on to final packaging. A number of storage stations (SS) are provided for IBC storage between process operations and queuing for process steps.

The advantage of this system is twofold: firstly the materials or product are totally protected from the operator and the environment, and secondly materials and the transportation equipment do not enter the process room, thus minimizing room sizes. Each process machine, the room and corresponding IBC discharge and receiving station is sometimes referred to as an island of automation.

232

Two types of process machine are located on the transportation floor: an IBC blending machine which blends the product materials in the IBC by rotation, and an IBC washing machine where IBCs are washed and dried.

Weigh stations are provided to verify complete emptying of IBCs after each discharge operation to a process machine. The verification and identification of the IBC and its contents at all the process operations throughout the first floor are carried out using transponder IBC tags or a bar code identification system.

Configuration

The layout configuration of the system is multi-variable allowing remote location of process operations. To achieve an operable AGV route configuration on the first floor requires development of an integrated three-dimensional design between ground floor process room layout and first floor discharge and receiving station. The ground floor process room layout affects the first floor AGV system design and therefore must be designed integrally. A typical AGV route configuration for a two-floor concept is shown in Figure 12.3 which shows an AGV route serving both discharge and receiving stations on both sides of a narrow aisle. Most AGV systems can be expanded to serve an automated warehouse: supply of raw materials to the dispensing area, packaging materials, and final storage of finished packaged pharmaceutical product.

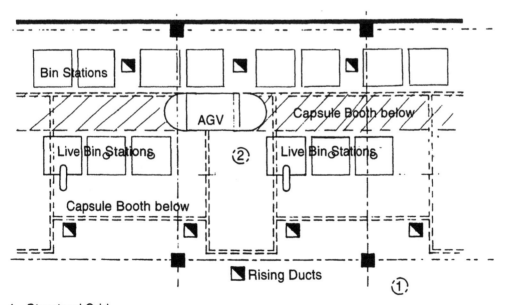

1. Structural Grid
2. Rooms between Process Booths

Figure 12.3 AGV route configuration

Automated Transportation

All movements of materials within a production facility can be achieved using automated transportation systems between the warehouse, process areas, packaging and shipping areas.

A number of solutions exist, some more applicable and complex to this type of facility than others. For example:

- automated guided vehicle system
- overhead monorail system
- transfer and shuttle cars
- conveyors.

The AGV is very suitable for carrying and transferring unit loads. It can be controlled automatically, receiving guidance and transport orders from its own land-based transport computer control system.

Vehicle guidance generally uses an inductive wire embedded in the floor coupled with radio or infrared communication between the AGV and a computer system.

More sophisticated systems utilize intelligent vehicle pre-programmed routes and dead reckoning with positional cross checking via three-dimensional special tri-angulation using laser scanning. The real time operation of the automated transportation system allows products and packaging components to be delivered where and when required on a just in time (JIT) basis, eliminating the need for intermediate storage and staging areas and thus generating improved process stage flow and material tracking.

Standardized Containers

The development of standard IBC sizes is an important part of any automated system. All packaged materials in bags or drums in a production facility are replaced with rigid construction sealed product IBC of specialized design dependent on application and the materials to be contained. Both the internal and external IBC designs have a critical effect on the operation of the plant. The IBCs are generally designed to blend dry powders and granules and to store compressed tablets and filled capsules between process operations. Other IBCs of a simpler design could be utilized to carry empty capsules, packaging materials, etc. The internal design of a powder IBC will depend on the material characteristics such as particle size, bulk density, flowability, etc. of the solids. For a tablet IBC, integral design would be influenced by product characteristics such as friability, hardness, etc. in order to reduce the potential for product damage to an absolute minimum. All types of IBC are enclosed in an external frame of identical dimensions which ensures exact spatial positions of the IBC inlet and outlet valves.

By standardizing the overall IBC dimensions, simplification of the automated transportation system, transfer and positioning, sealing systems at discharge and receiving stations, and the design of the washing machines and blending machine is achieved.

The IBC capacity is optimized to reduce the number of sizes, and consequently the number of transfer operations and formulations, weighing, etc. which, in turn, reduces the demand frequency on the transportation system.

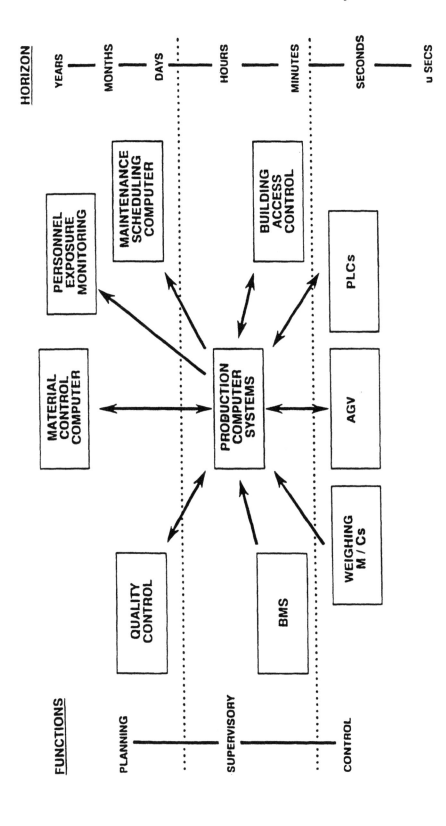

Figure 12.4 Process management and control. Key: BMS, building management system; M/C, machine; AGV, automated guided vehicle; PLC, programmable logic controller

235

Each production machine in the facility located is linked to the supply IBC by a discharge chute and a pneumatic vacuum return pipe to the IBCs at the discharge and receiving stations respectively. Both types are delivered and transferred by the AGV to the station where it is aligned and positioned with the corresponding link to the process machine.

Once positioned the sealing systems are activated and provide an efficient seal between the station discharge chute and the discharge flanges ensuring a clean process environment. Identical operations occur at the adjacent receiving station.

Depending on the level of machine automation and control capability the feed and return systems can operate unmanned over 24 hours ('lights out').

In many cases a complete island of automation is not always possible and manual operator intervention from a process room computer terminal may be required.

It is critical that complete sealing within the process supply and return loop is achieved at each station connection to prevent cross-contamination between areas.

Computer Integrated Manufacture

The manufacturing process and integration of the transportation and materials handling systems require a high level of computer integrated control and communication.

PLCs are provided at all materials handling stations including the washing and blending machines and the automated warehouse, to manage and control each individual operation. The manufacturing rooms are provided with local dedicated PLC units responsible for process control and monitoring.

The production host computer, which handles process monitoring and control production computer systems (PCS) is responsible for complete management of all facets of production and would directly control and co-ordinate information from all low level systems and PLCs. The PCS would not only control and monitor process operations but would also plan and implement production requirements, store production data, set up batch processing programmes and track process material by mapping and identification/verification at each process step.

A typical systems control architecture is shown in Figure 12.4, which shows the main functions such as:

* planning
* supervisory
* automated control.

Equipment Design Standardized Containers – Product IBCs

Process materials in various stages of manufacture from formulated batches of dry powder to the finished dosage form are stored and transported in stainless steel product IBCs of varying design. These IBCs are designed (Figure 12.5) with identical rectangular space frame dimensions and inlet and outlet valves that allow precise transfer and positioning at the discharge and receiving stations respectively.

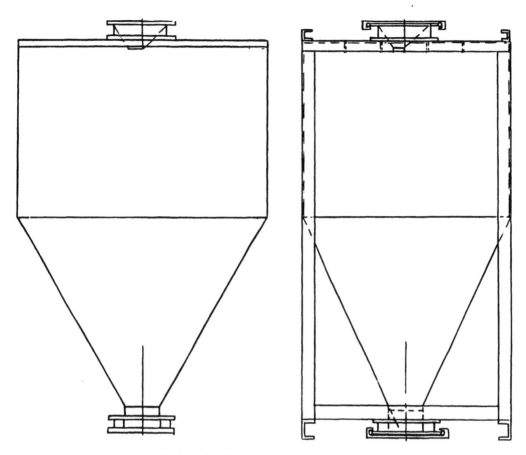

Figure 12.5   Basic standardized product IBCs

The product container's internal design varies depending on the type and character-istics of product material and the process operation to be carried out. The IBCs are built with specially designed inlet and outlet valves and slide cover plates to ensure maximum containment. These valves are automatically actuated by location mecha-nisms that ensure sealing from the environment. The IBCs can be considered as a mini process environment that must be resistant to contamination from the environ-ment and also does not contaminate the environment by leakage during transfer operations.

IBCs require cleaning before change of product to eliminate contamination of materials; this is accomplished using an automated washing facility. One advantage of the automated process is that this operation can take place outside normal oper-ating hours.

The application of Good Manufacturing Practice (GMP) to the design and fabri-cation of the IBC ensures that when in use product materials will not be trapped and held in crevices, joints, etc., and that process materials are not degraded or contaminated from other manufacturing materials.

The basic design of IBCs is discussed in the following sections.

Powder IBCs

Powder IBCs are used to transport, store and blend dry powders and processed granules (Figure 12.5). The design ensures efficient blending of its contents by rotating the IBC at a fixed speed. It is also designed as a mass flow hopper to ensure that most materials will discharge without the use of a vibratory system. Its capacity is optimized to allow a range of batch sizes to be carried and efficiently blended.

A typical powder IBC of 2000 litre capacity is able to hold and blend 50–850 kg of material efficiently at between 10 and 20 rpm. A comprehensive study of all blending operations needs to be conducted. It may be that other sizes will be required but the objective should be to limit alternatives as much as possible. More than one size increases capital costs, increases the complexity of the manufacturing operation and can affect the cleaning programme.

The frame size will directly affect transportation route widths and the overall height will effect the layout and building costs.

Tablet IBCs

Tablet bins (Figure 12.6) have a unique internal design for carrying easily damaged materials such as compressed tablets and filled capsule dosage forms, where product characteristics such as friability, hardness, size, shape, etc. are all critical. These

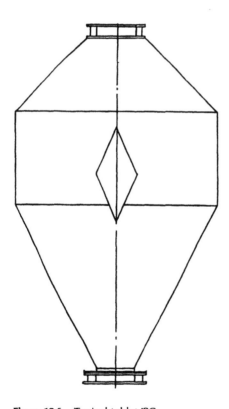
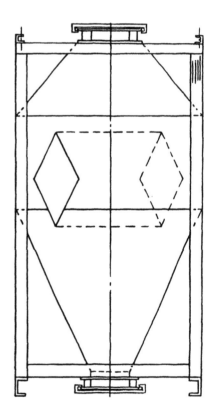

Figure 12.6   Typical tablet IBC

designs can be provided with removable spiral or baffle plate sections that reduce the tablet's kinetic energy as it enters the IBC and drops from the inlet, and also reduce the hydrostatic pressure which tends to crush the tablet as the head of material builds up. This design ensures smooth discharge of the dosage form without residual material being left behind.

The internal designs are complex but each IBC must be easily cleaned and dried in the washing facilities. IBCs with a simpler internal design to that of the tablet IBC are utilized for carrying empty capsules, and packaging materials such as bottles and caps, etc.

Various inlet and outlet valve configurations that take account of the physical characteristics of the material can be designed; these are activated by the standard system mechanisms.

IBC Discharge and Receiving Stations

Automated discharge and receiving stations are positioned for IBCs to discharge their materials to the process machines on the ground floor and receive processed materials through the pneumatic vacuum transfer system. Identical stations are used throughout the materials handling system with compatable transfer systems at the washing and blending machines. An example is shown in Figure 12.7.

Effective sealing must be maintained at all times to ensure elimination of:

- Process material leakage into the first floor technical and ground floor process room environments.
- Contamination of the technical areas from residual process materials remaining on valve faces when the IBC is removed from the station.
- Cross contamination of the process room atmosphere from the first floor technical area environment.

Each station is provided with the following main components:

- Floor mounted frame.
- Flanged sealing system.
- IBC transfer and positioning system.
- PLC control system.

Transfer and positioning of the IBC to and from the AGV to the station can be achieved in a number of ways such as an extending fork or roller conveyor. The latter method is preferred because it eliminates vertical movements, making the positioning, actuation of sealing and valve mechanism design less complicated.

Recent developments have allowed the AGV to drive the station roller conveyor transfer system by using a frictional slave drive. This reduces costs significantly by removing the need for roller conveyor drive units on every station used or for IBC transfer.

A variety of seal compression and inflatable designs and configurations are available. These include air blast purging on disengagement to ensure removal of all residual material; and both IBC and station flanges are provided with flange slide cover plates to ensure complete containment of any residual process materials.

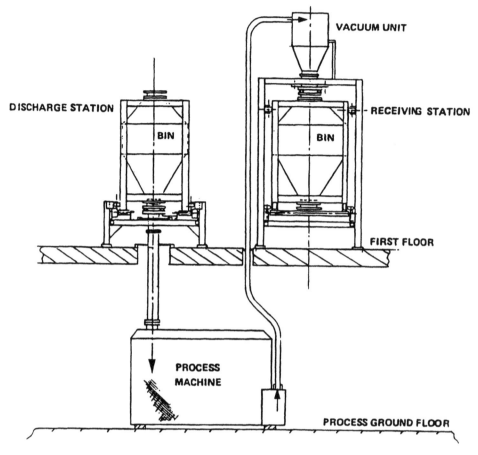

Figure 12.7 Process configuration

The control of each station's sequential operations, fault monitoring and communications to and from the AGV is provided from a dedicated PLC.

The design of both stations is very similar.

Automated Transportation System

The type of system chosen will be dependent on a number of parameters:

- process design
- level of automation required
- costs
- system flexibility.

As mentioned earlier the available systems are:

- fixed conveyors
- pneumatic transfer

240

- shuttle cars
- transfer cars
- overhead monorail systems.

Automated Guided Vehicle Systems

It is not the intention to discuss the merits of each but to give an overview of the more common and latest design systems that have been adopted in the more recently built pharmaceutical facilities. Both overhead monorail systems and automated guided vehicle systems provide the capability of a fully automated transportation system able to link every process stage. The OMS rail system has the disadvantage that it must be very strong structurally to be capable of carrying loads of several tonnes. The AGV system has generally been adopted as the preferred system.

The AGV is an unmanned computer controlled vehicle capable of providing IBC transport and guidance to and from process station operations and the warehouse (process here includes packaging). The number of AGV vehicles required is dependent on the movement demand which depends on the process activities.

Two main types of guidance systems for AGVs are available. Firstly the AGV can follow an inductive wire set into a concrete floor. In this case optical or high frequency radio floor codes are used for the station positioning and intersection control which are recognized by the AGV on-board intelligence.

Alternatively AGV dead reckoning (by wheel movement monitoring) guidance can be employed for the complete system controlled by the pre-programmed on-board intelligence. The AGV position is continually cross checked and verified using three-dimensional laser scanning from the AGV to strategically positioned bar codes.

This second method is more sophisticated but the two systems can be integrated using wire guidance for straight sections with dead reckoning off wire travel for short route segments outside the wire guidance system, without the need for cross-checking.

The computer permits simple reprogramming to modify the guidance path or vehicle route for special tasks, process route modification or system expansion. The AGV can communicate by transmission through a photocell at the stations, or more preferably, by using infrared or radio transmission.

To confirm that all the correct positions are reached during IBC transfer, hand-shake communications between the AGV and the discharge and receiving stations are carried out by local photo-optical or infrared systems.

Product IBC Blending

A critical step in the manufacture of any powder based product is the ability to blend material at various stages of the process using the product container, in this case the IBC. The IBC is rotated in a blending machine at a fixed speed which depends on the physical characteristics of the mixture, and for a preset number of rotations for each batch of material. This blending procedure needs to be validated for each product as the speed and number of rotations may vary in relation to the particle characteristics of the mixture.

Because the process materials remain in the IBC during the blending operation, this operation can be located in a technical area rather than a process room and this, in turn, minimizes HVAC and floor area costs.

The blending machine is equipped with a rotating cage to hold and lock the IBC in position. There is also a safety fence to protect operators during rotation. IBCs are automatically transferred and loaded into the blender cage from a transfer conveyor in the adjacent transport aisle.

Each programme in the local PLC represents the required number of rotations for each product and any unique requirements; typically 30 programmes are provided.

The PLC monitors all sequential operations during transfer and blending operations, verifying that each step has been achieved before proceedings, and indicating faults immediately to ensure a safe and reliable operation.

IBC Washing Machine

Washing machines are required to clean, wash and dry all types of IBCs. They are delivered by AGV to the washing area where they are washed and dried before collection by the AGV for storage.

The washing machine design is a complex but a necessary part of the process system to allow various process materials and mixtures to be manufactured. It can be likened to a clean-in-place system. Basically it is a stainless steel cubicle which allows a dirty IBC to enter and a clean IBC to leave. Once inside the washing cubicle high pressure spray heads enter the top and bottom of the IBC through the open valve apertures. Depending on the nature of the product materials, various washing procedures and formulations of the washing medium may be required. Some product materials are very hydrophobic and special conditions may apply. Usually high pressure hot water is all that is required.

Pneumatically operated actuators open and close the inlet and outlet slide cover plates and the butterfly valves to allow spray head access. They can be cyclically oscillated during the washing cycle to ensure complete cleaning of the valve. The exterior is washed using fixed high pressure spray jets positioned inside the cubicle. An air drying system is provided to completely dry the IBC inside and out after washing and rinsing has been completed.

Careful washing machine design is essential to ensure that all IBC designs including internal baffles, valves and slide plates are washed leaving no residue. Various grades of water, e.g. Purified Water BP, may be necessary to meet GMP requirements.

A totally integrated system for solid dosage manufacture (tablets and hard shell capsules) based on the principles outlined is illustrated in Figure 12.8.

Each numbered unit operation in Figure 12.8 is explained as follows:

(1) Bag/drum emptying hopper. With sieving in line and discharge of the sieved powder into an IBC.
(2) Product dosing by screw and/or vibrating feeder from IBC to IBC.
(3) Product dosing by screw and/or vibrating feeders from silos to IBC.
(4) IBCs handled by: hoist, transpallet, fork lift and AGV.
(5) Direct feeding of a process unit granulator by IBC (gravity or vacuum systems).

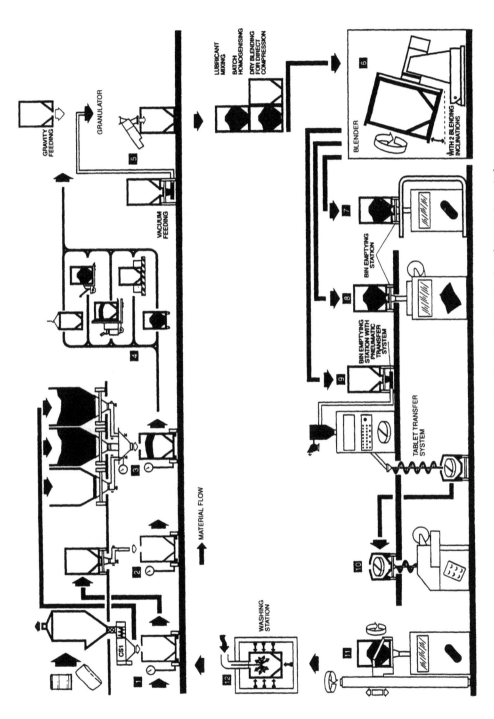

Figure 12.8 Integrated manufacturing system based on IBCs (courtesy of Zanchetti and CSrL, Marcheschi, Vittoriano, Italy)

(6) IBC tumbling by means of a rotating blender.

(7) Direct gravity feeding of a process machine by an IBC discharge station at the same level.

(8) Direct gravity feeding of a process machine by an IBC discharge station from a second level.

(9) Vacuum feeding of a process machine from an IBC discharge station and unloading of the final product directly into a tablet IBC by means of a vibrating column system.

(10) Direct gravity feeding of a blister machine by a tablet IBC and vibrating column system.

(11) Lifter pillar to blend and/or feed directly a process machine by an IBC.

(12) IBC washing and drying station, simplified semi-automatic or fully automatic models, static and rotating station, IBC loading by transpallet, fork lift or AGV system. Essentially a clean-in-place system.

A comparison between Figures 12.8 and 12.9 will show how the traditional methods of handling materials (Figure 12.9) for a tablet coating process have been largely superceded by the more sophisticated use of IBCs.

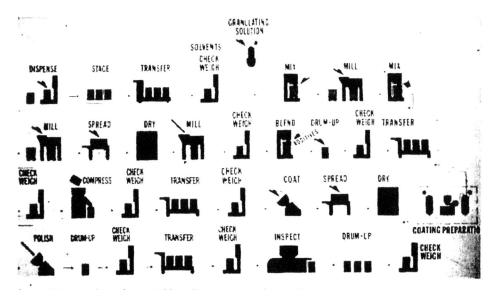

Figure 12.9 Traditional material handling operation for a tablet coating process

13

Flexible Manufacturing Systems and Automation

'Get to your places!' shouted the Queen in a voice of thunder, and people began running about in all directions, tumbling up against each other: however, they got settled down in a minute or two, and the game began.

The Changing Business Environment

As competitors from all over the world move into local markets, the local environment becomes less significant, and worldwide standards of performance start to apply. Any industrial company that hopes to retain an appreciable share of its market now faces an increasing challenge to compete worldwide.

Manufacturing industry must meet a number of challenges. It must:

- raise standards of product quality to close to 100%
- introduce new and improved products in shorter development times
- attain both higher productivity and reduced costs
- become flexible enough to adapt to fast and unexpected changes
- make available more and better information for the company and its customers.

Business Challenges

If there were no other reason for re-thinking the way the company operates, the drive for 100% quality in goods and services would in itself be sufficient incentive.

Total, or 100%, quality and reliability is fast becoming the norm in the international market place. Companies expect it from their suppliers, and expect it with every product they use. Achieving 100% quality in a product requires 100% quality in the way it was designed, planned, and built. Continuous control of quality is a feature of the process industries; out of necessity it is now being adopted by all sections of manufacturing. Automation with its consistent repeatability, is often the only way to achieve it.

With changes in the marketplace, government regulations, and emerging technologies, as well as the cost and availability of drugs, raw materials and energy, changes to the product line, and alterations to the product, the process must be fast, painless, and inexpensive. The speed with which a company can react to change depends largely upon how well its business functions communicate.

With all the above challenges, better information is needed internally to manage the business. Information is needed to meet the company's legal, contractual, and fiscal obligations; and to satisfy government obligations, national and local, which need an increasing variety of data for statistical and regulatory purposes.

Meeting the Challenge

The challenging business environment is putting pressure on each business function within a company. And each function is trying to meet the challenges by:

- Bringing in as many of the technological innovations as it can convince management to invest in, such as, computer aided engineering and flexible manufacturing systems.

- Adopting new techniques and business methods, like just-in-time manufacturing, group technology, and cooperative partnerships with suppliers and subcontractors.

The pharmaceutical industry tends to concentrate its resources in the search for new unique therapeutically active compounds; however, the pressures on manufacturing costs are forcing the industry into engineering development, for example, where the emphasis is on reducing the production cycle, and building in product quality from concept through to release. Many companies are starting to design for automation, simultaneously taking into account in their designs the product, the process, and the production facilities. Operations and maintenance engineers take part from the earliest phase of the product design.

In brief, each function is managing to achieve improved productivity and reduced costs, often applying an unplanned combination of innovations, in accordance with the latest developments in their fields. The net result is what has become islands of automation.

But no random combination of these innovations will be sufficient to achieve success in today's marketplace. The information systems that support them operate on different pieces of equipment from different suppliers, and they are distributed throughout the plant. They are also generally implemented as stand-alone or departmental systems and were not designed to work together. Individually they can provide significant benefits to a company, but their inability to work together limits their contribution to the company's overall competitiveness.

The installation of independent systems has, until recently, been the natural, and in fact the only, way for industrial companies to implement the new technologies, techniques, and relationships, in a more strategic way: one that ensures that they not only work together, but work together to meet the company objectives. In other words, an industrial company needs a strategy for computer integrated, rather than computer aided manufacturing.

Computer Integrated Manufacturing

Computer integrated manufacturing (CIM) is a business strategy, rather than an automation or computer strategy. Its aim is the success of the company as a business entity, not simply the improvement of the company's individual engineering, production, and administrative functions. For integration to be effective, the business must be viewed as a whole, unified system, rather than a collection of individual functions.

The term computer integrated manufacturing means more than just the production side of the business. A full interpretation of the acronym CIM might be computer integrated management of the business.

Top Down Plan (Figure 13.1)

A departmental-only system was the natural approach in the past when users needed to get their feet wet in the new technologies. Today, a top down, or total company, view is required to provide the integration needed to maintain competitiveness. Because this approach affects the whole business, and because of conflicting departmental objectives, it must be led by the company's executive management.

The key to integration is information; the information created, maintained, and disseminated by each area of the business. The information is held in many different forms, and, where automated, may be on different types of system.

The pharmaceutical industry is unique in the procedures and methods of manufacture that are used to ensure the integrity of its product. These are essentially achieved by three main functions: quality assurance (QA), current good manufacturing practice (cGMP), and quality control (QC). To avoid confusion and to clarify their interrelationships these terms are defined here:

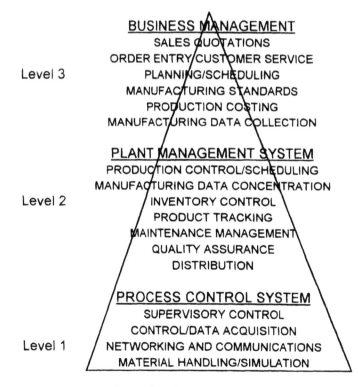

Figure 13.1 Computer integrated manufacturing

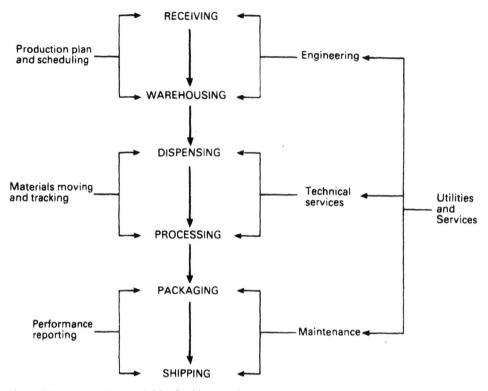

Figure 13.2 Integration model for flexible manufacturing

Quality assurance is the sum total of the organized arrangements made with the object of ensuring that products will be of the quality required by their intended use. It is cGMP plus factors such as original product design and development.

Good manufacturing practice is that part of QA aimed at ensuring that products are consistently manufactured to a quality appropriate to their intended use. It is thus concerned with both manufacturing and QC procedures.

Quality control is that part of cGMP which is concerned with sampling, specification, and testing, and with the organization, documentation, and release procedures which ensure that the necessary and relevant tests are, in fact, carried out, and that materials are not released for use, nor products released for sale or supply, until their quality has been judged to be satisfactory.

This necessarily generates considerable documentation and record storage. It follows that any automation system must maintain these requirements as a minimum. The principle is, therefore:

> There should be a comprehensive system, so designed, documented, implemented, and controlled, and so furnished with personnel, equipment, and other resources as to provide assurance that products will be consistently of a quality appropriate to their intended use. The attainment of this quality objective requires the involvement and commitment of all concerned, at all states. (*The Orange Guide*, 1983)

More than 13 000 robots are currently in use in the United States, and Japan reportedly has more than 400 000. Virtually every major US pharmaceutical firm has at least one robot on site (Kanig and Rudic, 1986). Productivity has risen between 10 and 35% and the cost reduced from $26 per hour for an operator to $6 per hour for a robot.

The model shown in Figure 13.2 illustrates the main dimensions in pharmaceutical production. In addition, all parts of this model require the input of energy, and the central core is overseen by Quality Control.

There are two sets of employees in this model, the operations personnel and the support personnel.

Receiving – Warehousing – Dispensing

One of the problems in the pharmaceutical industry is the relatively small volume of the components and the large number of different types of the same component, e.g. labels in different languages. Investment has to be justified on the following economic basis:

- Savings in manpower.
- Improvements in materials handling.
- Security.
- Improvements in efficiency in production.
- Planning and scheduling.
- Improvements in inventory control.

Generally, in the early 1970s, to make a system viable 10 000 storage locations were required. This was due to the high cost of automated guided vehicles (AGVs).

The scheme described here illustrates the series of events from the arrival of the item at the warehouse of the manufacturing facility and its dispatch as goods for sale.

Delivery of raw materials can range from discharge from a bulk tanker to a few sacks, drums, bottles, or packaging components ranging in weight from a few kg to over 1000 kg. Where a system of AGVs is used, then a holding and collation area is necessary. This will necessitate some repackaging operations. Generally, special metal constructed pallets are required. It may also require a 'skinning' operation to secure the packages or containers during the transfer to their storage location in the 'in-house' pallet. The repackaging is a manual handling operation. It is very difficult to arrange delivery of all materials in a standardized container, and at this stage of development in the industry it is easier and cheaper to modify the package on receipt.

It is also necessary on receiving the goods to generate the necessary documentation which identifies the product and initiates the in-house services such as quality control (QC). Initially, the material will be stored in a quarantine area awaiting QC clearance. When the product has been either accepted or rejected the material can be moved onto the company pallet, secured, modified, the storage location defined, and the pallet transferred, using the AGV. The human element is removed and errors in labelling or storage location reduced. The inventory is updated at this stage, and the material becomes available for the production process.

In the alternative scenario, which is more common, material is requisitioned from the warehouse and transferred to the dispensary for weighing and collation into batch loads. Any residual material is returned to the warehouse stock. A system can be devised that automatically subtracts the weights from the inventory at the dispensing point and activates the re-order procedure when a critical stock control level is reached.

This enables the inventory levels to be maintained at the appropriate level. What is probably more important in the pharmaceutical industry is the accuracy of these stock levels, owing to the high value of drugs. The final package containing the dosage has many components, and the situation is further complicated by different packaging elements being required for different export markets.

There are a number of ways of identifying materials that can be read by photo-sensing devices; bar codes are one, metal tags and small electric storage devices such as transponders are others. These can be scanned quickly and easily and the information retained in a central computer memory.

Processing

This part of the overall process requires monitoring by personnel at various stages of the manufacturing cycle. It is essentially a checking and data logging function. There are two types of processing operation:

- semi-continuous batch processing (mechanical make-up)
- batch processing (manual make-up).

In the first case, material is dispensed from the storage silos into a holding vessel and identified as either a discrete batch or sub-batch. Sub-batches are used where a final blending operation takes place to form the discrete batch. Which path is chosen will depend on the manufacturing operation that is the rate determining step in the overall process.

Generally, once material is dispensed from the silo it is totally enclosed and moved by pneumatic or other material handling devices to each process step. It is essential that at critical stages of the process, material balances are measured and recorded. This ensures that each blender or drier, etc., is fully discharged before accepting the next batch or sub-batch.

Tight control is necessary, especially when the dosage form has completed the manufacturing cycle.

When the materials are dispensed into a container which constitutes part of the manufacturing operation, checks on the overall balance are usually made at the end of a series of operations, e.g. when the tablet is compressed or when the bulk liquid or ointment is formed and stored ready for packaging. In some cases, an in-process storage area is required, for example, where tablets are required to be coated or sterile products are awaiting the results of sterility testing. In some cases, batches of material can stay in the in-process area for several weeks. In all integrated designed plants this area can be part of the warehouse facility.

Each batch must have a batch manufacturing record document on which the following information is recorded at the time each action is taken:

1. The batch identifying number of each of the starting materials used.

2. Where the master formula permits variation in the quantity of starting material, a record of the amount actually used.

3. Dates of commencement and completion of manufacture and of significant intermediate stages.

4. Where more than one batch of a given starting material is used, a record of the actual amount of each batch.

5. The batch identifying number and amount of any recovered or re-work material added.

6. The initials of the person(s) who weighed or measured each material and the initials of the person(s) who checked each of these operations, this check being not only of the quantity, but also of the labelled identity and batch number of the material. The need for the second series of check initials may diminish if equally effective electronic and other supporting systems are in operation.

7. The amount of product obtained at pertinent intermediate stages of manufacturing.

8. The initials of the person responsible for each critical stage of manufacture.

9. The results of all in-process controls, with the initials of the person(s) carrying them out.

10. Reference to the precise items of major equipment used, where several of the same type are available for use (i.e. where equipment is replicated). This information may be recorded in 'plant usage logs'.

11. Details of, and signed authorization for, any deviation from the master formula and method.

12. The final batch yield and the number of bulk containers.

13. Signed agreement by the process supervisor that apart from any deviation noted as in (11) above, manufacture has proceeded in accordance with the master formula and method, and that process or yield variations are adequately explained.

These records have to be retained for five years or longer, depending on the availability of the product.

Another important aspect of the function of the dispensing computer is the ability to store a series of formulations and a list of ingredients as specified by the batch manufacturing record (BMR). Only when the correct identification number is fed into the dispensary weighing system can the correct formulation be recognized. As each material is weighed and its weight recorded a self-adhesive label will be created identifying the product, batch number or product, dates and batch number of the material. This is attached to the BMR and signed for by two operators.

Packaging – Warehousing – Dispatch

Packaging operations in the pharmaceutical industry are complex and require attention to detail. For solid dosage forms such as capsules and tablets, the batch is delivered to the packaging area as a complete batch. The individual dosage form is then subjected to a series of operations to ensure that the correct number of items are filled into the container or that the blister pack is completely filled. The regulatory authorities' records show that there are more problems associated with packaging operations and more recalls due to labelling errors than any other.

The packaging equipment is built by specialist manufacturers, and each machine will require upwards of three operators, depending on the diversity of the operation and the extent of the labelling and cartoning. In a number of pharmaceutical companies some of these operations such as bottle orientation, capping, labelling, and collation (bundling) have been replaced by robots.

At the labelling station an operator is responsible for checking that the bottles are capped, correctly labelled, and that any overprinting has been correctly applied at the rate of up to 100 units per minute. Using machine computerized vision developed in conjunction with the robot, this extremely tedious operation can be eliminated and product security can be improved (QA). One feature of this system is generation of batch documentation and a visual display of a running histogram of the packaging lines' productivity for the batch currently being filled and packed.

The objective here should be to provide a robot capable of performing a number of material handling functions and to provide the necessary software for checking that a number of operations have been successfully completed, including generation of batch documentation.

In Figure 13.3 the layout flow diagram shows the number of stages that are necessary in the preparation and filling of a sterile product.

Pharmaceutical companies operate their production facilities on a batch basis. Even companies such as Merck which have computer controlled and monitoring processes in secondary production need to be able to identify specific batches of products.

In the case of these processes, large quantities of new materials will be transferred, probably by fork lift truck, to a silo storage location once a week (or month).

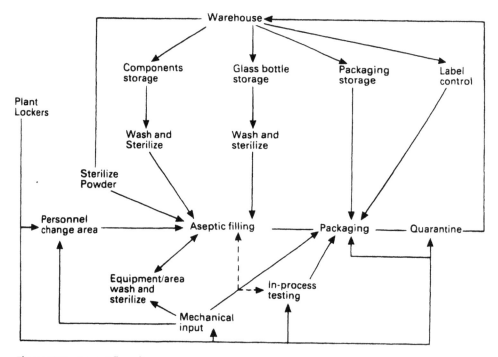

Figure 13.3   Layout flow diagram

The frequency depends on quantity and usage. These are known as dump stations, and here the material is removed from its primary package into the storage silo. From here it will be dispensed automatically by a pneumatic transfer system and load cells to achieve a material balance. The material is effectively removed from the inventory at the point of leaving the warehouse, and no material is returned.

Once the product has been successfully packed and collated into an outer carton containing say 100 unit packs, the carton is transported to the warehouse. This is achieved by using a conveyor and by affixing a bar code onto the outside of the carton which is read by a photo-sensing device. The product can be located and stored on a pallet until that pallet has been filled. It is then moved into its selected storage location by an AGV.

The product is now available for sale and dispatch, and is entered into the company inventory for finished stock.

Summary

For any flexible manufacturing system the following information will be required:

1. Process flow diagram.
2. Definition of the flow rates.
3. The impact of the plant building layout.
4. Type of transfer system recommended.

The transfer system can be divided into a number of different options. These include conveyors which may be overhead and used by a powered monorail system or a ground mounted conveyor system. They are simple, reliable, and reprogrammable. They are also robust and have a high capacity. Against this must be their rigid layout, and they also have restricted access. The overhead conveyor will also require considerable supported infrastructure.

Computer Control

Since 1970 the explosion in the development of the microprocessor, the programmable controller, and personal computers has provided tools to make substantial productivity improvements in manufacturing systems. Many of the major pharmaceutical products in the oral solid dosage field are now produced by automated plants. Merck's ALDOMET manufacturing facility is well known and well documented (Lumsden, 1982).

The purpose here is to consider some of the ideas incorporated into these plants and to look at what has been achieved with products that are not produced in large volumes and, therefore, require a more flexible approach. The key driving forces towards providing a better utilization of these assets are:

- The rising cost of labour relative to productivity increases.
- The high cost of energy and raw materials.
- Poor material handling facilities.
- The high cost of quality control.

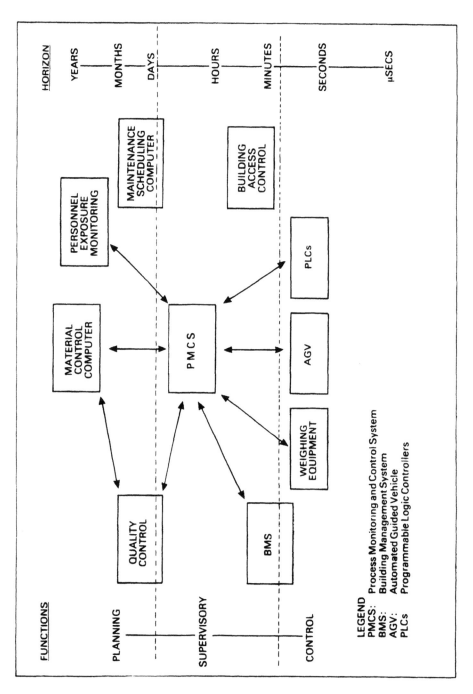

Figure 13.4 Systems architecture

- Inefficient use of manufacturing equipment.
- Short runs and a variable product mix.

It has been suggested that, over the next decade, in any productivity increase almost 60% will be provided by new and existing technology, 25% will be provided by capital and 13% by labour management (Morley, 1984).

All of these pressures are dependent on control. It is the control of the process, inventory, information flow, materials handling, and the utilization of plant which is the essential part of automation.

The facility requirements are:

- Data logging and process control of various parameters, tablet weight, hardness, thickness, etc.
- Mass balances.
- Batch control for various recipes and sequences.
- Optimization of batch control parameters.
- Flexibility to change the process, size of system, etc.
- Local operator interface in hazardous area.
- Interface to field devices suitably protected against hazardous area (intrinsically safe circuits, etc.).
- Operator control/display within the control room via VDU/keyboard/printer devices.
- Calculations and optimization of sampled and/or measured variables for analysis.
- Printouts and logs.
- A relatively small computer system.

Generally, to do this efficiently requires the use of a microprocessor or computer. Depending on the complexity of the operation, the number of parameters to be recorded, and the degree of control required, there are a number of systems that can be used. The interactions of the systems are shown in Figure 13.4 and the flow of information in Figure 13.5.

Systems

In the application of computer control to a plant, three main types may be used:

- A data acquisition system.
- A distributed control system.
- A centralized computer system.

The following describes briefly each system and highlights those points which should be considered when evaluating the requirements, and some suggestions/conclusions on the choice of a possible system.

This is not an exhaustive list for selecting a system, but more a discussion to assist in making some basic decisions.

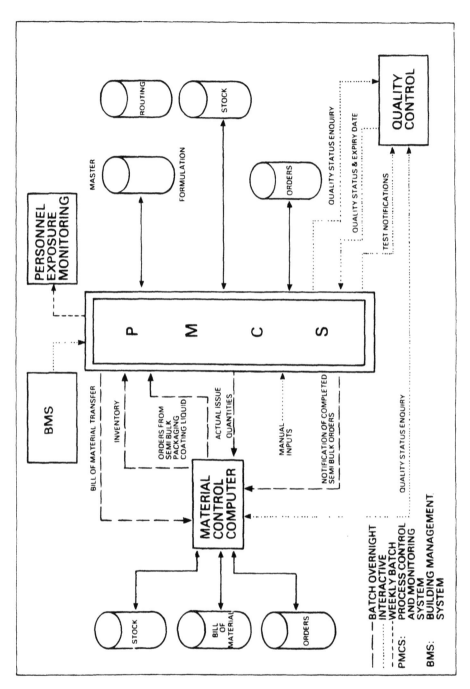

Figure 13.5 Data flow

Figure 13.6 shows the layout of a data acquisition system. They are small devices for low levels of inputs only. Some have limited computing power, and can be used to produce a printout of the information gathered. No plant control is provided, and only a limited amount of computing power is available.

Some devices can provide a serial type link to other computers like a Vax, which could then be used to store data or manipulate the data in some way.

The field inputs must be cabled back to a control location and the system interface would require safety protection.

Advantages

- Small and cheap.

Disadvantages

- No control provided.
- Limited data processing.
- Limited number of inputs.
- No data storage.
- Long lengths of field cabling required.
- No local operator interface.

Figure 13.7 illustrates the layout of a distributed control system (DCS). A distributed system means that not only are the hardware items geographically distributed, but the functions such as processing, power, and the software are also distributed.

Each hardware item performs a specific task, and information is passed between the various items via a dual data highway. This highway provides good security for communications.

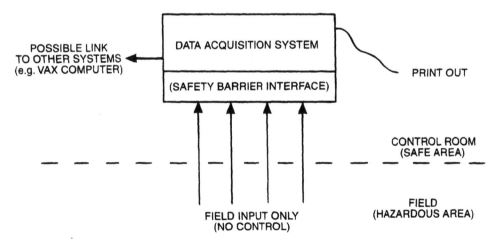

Figure 13.6 Data acquisition system

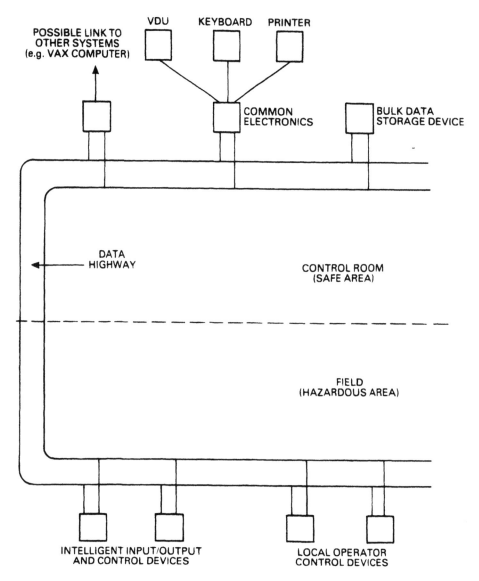

Figure 13.7 Distributed control system

The field devices, such as the input, output, and control devices, must be protected by suitable housing and safety equipment for use in a hazardous area.

In this type of system, the computer type functions are distributed throughout, using smaller microprocessor devices.

DCS systems can support local operator control devices located in the plant area, provided that they are adequately protected.

Process control can be provided for continuous variables such as pressure and temperature and can also provide batch or sequence control. It can be linked to other devices such as an existing Vax computer.

Advantages

- Batch, sequence, and continuous control are available.
- Data logging and data storage capacity.
- Calculation on the stored data can be performed.
- Logs and reports can be generated.
- Local operator interfaces can be provided.
- Batch and sequence control can store and use many different recipes and values, and it can be used to optimize process control.
- Easily expandable and very flexible.
- Short field cable lengths.
- High security of control as all functions are distributed.

Disadvantages

- System more suitable for plants with greater than 500 loops.
- Greater cost due to its large capacity.
- Local plant equipment must be protected in safe enclosures.

Figure 13.8 shows the layout of a centralized computer control system (CCS). This is very similar to the distributed control system discussed earlier. It can perform all the same functions, but they are centralized within a computer system. It is normally the process control manufacturer's standard, but other computers can be used.

Advantages

- Same as distributed system.
- System size tends to be intermediate between the systems illustrated in Figures 13.6 and 13.7.

Disadvantages

- Long lengths of field cabling required.
- System operation dependent on one device (the computer), thus a redundant or back-up computer may be required to increase reliability.

The type of system required should provide the following facilities:

- Data logging and process control of various parameters, e.g. temperature, spray rates on coating operations.
- Batch control for various recipes and sequences.
- Optimization of batch control parameters.
- Flexibility to change the process, size of system, etc.

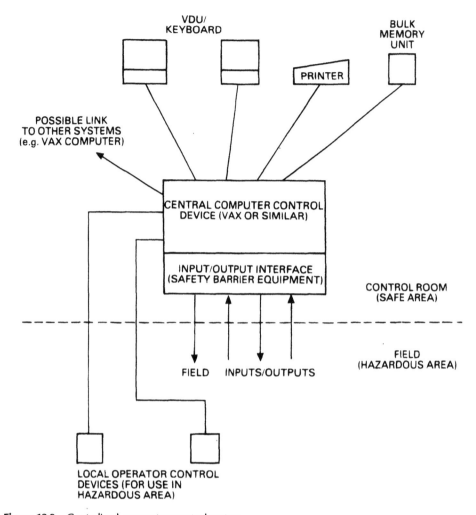

Figure 13.8 Centralized computer control system

- Local operator interface in hazardous area.
- Interface to field devices suitably protected against hazardous area (intrinsically safe circuits, etc).
- Operator control/display within the control room via VDU/keyboard/printer devices.
- Calculations and optimization of sampled and/or measured variables for analysis.
- Printouts and logs provided.
- A relatively small system (150 loops approximately).

With these requirements in mind either the centralized or distributed control system would be suitable for a facility.

The centralized system is preferable to the distributed system for the following reasons:

- It is smaller and cheaper.

- System input/output interface is simpler, and the distributed system requires special housings.

- It allows the possibility to use an existing computer.

Most process control and data logging manufacturers use their own computers, and their software structure may not be comparable with the in-house computer software. However, it is possible to provide the interface necessary, but this can cause problems.

Batch Control Models and Terminology

The following example of how the ISA S88.01 1995 standard can be applied to a pharmaceutical batch process is provided by Nigel Cole.

The overall purpose of this standard is to provide a common approach to defining and modelling batch processes and their associated controls.

With the growing use of increasingly complex computer systems in the manufacturing process, e.g:

- PLCs for plant control

- SCADA for plant control and monitoring

- MES for electronic work instructions and batch reporting

- MRP for scheduling, planning and management reporting,

it is becoming increasingly important to define processes and data in a structured and consistent manner. This makes the design and implementation of interfaces between these systems less complex, and reduces the scope for costly errors due to simple data translations or formats; the use of consistent terminology reduces the scope for confusion and reduces engineering time.

The remainder of this chapter outlines the use of the ISA S88 Batch Control Standard to define a pharmaceutical manufacturing facility. For this purpose a typical solid dosage manufacturing centre (SDMC) has been used. The aim is to illustrate that the ISA S88 standard is of direct relevance to the pharmaceutical industry.

Reference should be made to the ISA S88 standard to explain the terminology used in more detail. Only a brief description of the relevant parts of the standard is given here.

The ISA S88 standard defines a set of terminology for describing batch processes and then goes on to define a series of hierarchical models to structure the process, equipment and data. Each of the following sections describes briefly one of these models and illustrates how the SDMC could be defined using the model. The example does not detail the whole of the SDMC as this would involve much repetition; instead the granulation stage of the process is used to illustrate the more detailed lower levels of the model.

The hierarchical nature of the models makes it possible to define common activities once and just copy them where required. This allows for the standardization of common activities across the facility.

The ISA-S88.01 document goes on to describe other activities associated with batch control: exception handling, alarms, batch reports, sequence modes and states, etc. It also describes the relationships between the models defined. These topics have not been included in this overview.

Process Model

A general description of the tablet manufacturing process in a solid dosage manufacturing centre (SDMC) is given in Figure 13.9. This model is based on a typical SDMC.

The batch process can be organized into a hierarchical model. The batch process is described as 'a process that leads to the production of finite quantities of material (batches) by subjecting quantities of input material to a defined order of processing actions using one or more pieces of equipment.'

The ISA S88 process model is divided into four levels as shown in Figure 13.10. Each of the levels in the process model is described as follows:

Process: A sequence of chemical, physical or biological activities for the conversion, transport or storage of material or energy.

Process stage: Part of a process that usually operates independently from other process stages. It usually results in a planned sequence of chemical or physical changes in the material being processed.

Process operation: Each stage consists of an ordered set of one or more process operations. Process operations are the major processing activities that define requirements without considering the actual equipment required. A process operation results in a chemical or physical change in the material being processed.

Process actions: These describe the minor processing activities that are combined to make up each process operation.

The manufacturing process in the SDMC can be described in accordance with the four levels of the process model.

Following the structure down through each level is done in a logical manner. Some of the lower level components may be used in more than one of the higher level components. For example a process action may be defined to challenge a balance. This action could be used in more than one process operation, i.e. manual dispensing and coating solution preparation; the definition of the action is the same in both operations. In such cases the action will only be defined once and referred to in subsequent uses.

Process

Solid dosage manufacture: Applying the model to a secondary pharmaceutical plant the SDMC becomes the top level of the model. Other manufacturing centres within the manufacturing site would also be defined as processes in a model of all the manufacturing activities, e.g. liquid manufacture, aerosol manufacture.

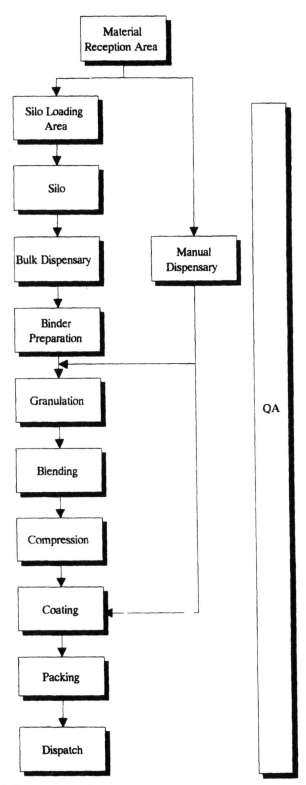

Figure 13.9 SDMC process overview

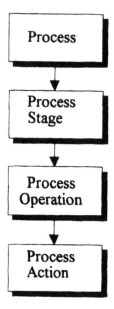

Figure 13.10 Process model

Process Stages

The next level of the model breaks the overall manufacturing process into self contained manufacturing activities:

- Raw materials handling
- Manual dispensing
- Bulk dispensing
- Binder preparation
- Granulation
- Blending
- Compression
- Coating
- Packing
- QA.

The individual process stages perform the following functions:

1. Raw materials handling
   Receipt and checking of raw materials.

2. Manual dispensing
   Smaller quantities of material are manually weighed out according to the batch recipe. A variety of different types of dispensary are used depending on the hazards associated with the material being dispensed.

3. Bulk dispensing

   Large quantities are weighed out using a semi-automatic dispensing system designed to handle larger quantities.

4. Binder preparation

   The agent used to wet the powder during granulation can be water, a solvent, or a solution made up to a specific recipe.

5. Granulation

   The dry contents of the bin are passed into a granulator and mixed with quantities of binder and/or solvent to produce a granule. This is passed through a wet mill into a fluid bed drier. After drying the granule is transferred through a dry mill to another bulk handling bin.

6. Blending

   The contents of the bulk bin are blended together in the bulk bin. This can occur before or after the granulation stage and may include the manual addition of additional material.

7. Compression

   The granules are fed into tablet presses.

8. Coating

   A solution is prepared and sprayed over the tablets.

9. Packing

   The tablets are transferred to the packing hall where they are packed into packaging for their market place. This could be individual cartons or bulk packs for transfer to overseas markets. For some markets the granule is packed for further processing in other countries.

10. QA

    This is continuous throughout the manufacturing process and could require the definition of specific operations for different stages of the process.

Process Operations

The process operations that make up each process stage are listed below. These are all the operations that could be used; any specific process stage will comprise one or more of these operations and operations may be repeated.

Raw materials handling
    Material arrives in SDMC
    QA
    Sieving
    Load silo
    Transfer to holding area
Manual dispensing
    Equipment check
    Material check
    Booth dispensing
    Isolator dispensing
    Reconciliation
    Sampling
    Sieving

Bulk dispensing
    Equipment check
    Material check
    Bulk dispense
    Transfer manually dispensed material
    Discharge to hopper
    Reconciliation
Binder preparation
    Water addition
    Solid addition
    Heat
    Slurry preparation
    Slurry addition
    Heat
    Agitate
Granulation
    Equipment check
    Load mixer
    Dry mix
    Solvent addition (binder etc.)
    Wet mix
    Mill/transfer to drier
    Drying
    Mill/discharge to hopper
    CIP
Blending
    Transfer to blend station
    Blend
    Transfer to holding area
Compression
    Equipment check
    Material check
    Manual tablet compression
    Compression run
    Process tests
    Layering press
    Discharge
    CIP
Packing
    Line clearance
    Material check
    Machine set up
    Material sorting
    Initial run
    Main run
    Carton
    Sampling
    Casing
    Overwrapping

Bulk packing
Granule packing.

Process Actions

Each of the process operations can be further divided into process actions. Process actions are the smallest description of process functionality and typically describe single control actions. These process actions can be specific process actions defined for the process equipment or they may be prompts for operator action essential for the correct operating procedure of the equipment being used. Throughout this process there are points at which operator identity must be logged and if there is a variation from the norm a supervisor confirmation of the action taken is also required; this should be more than simply entering a password, e.g. electronic swiping of an ID card, or use of bar-code reader. To keep the following list readable, these actions are assumed to be part of the definition of the specific process action they are confirming. Similarly event logging is assumed to be part of the Action definition. The division of process operations into process actions is illustrated in Figure 13.11 using the granulation operation as an example.

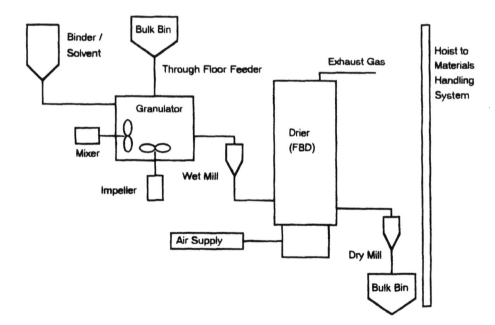

Figure 13.11   Diagram of a typical granulation facility

Process stage
    Process operation
        Process action
Granulation
    Equipment check
        Area clean
        Clothing challenge
        Ambient condition check
        Spray arm check
        Binder vessel level
        Prime binder pipe
        Spray arm position
        Pre-heat the FBD (drier)

    Load mixer
        Position hopper
        Open transfer value
        Transfer material
        Close transfer valve

    Dry mix
        Start impeller
        Start chopper
        Run timer

    Solvent addition (binder etc.)
        Start impeller
        Start chopper
        Start pump
        Run timer
        Stop pump

    Wet mix
        Start impeller
        Start chopper
        Run timer
        Stop impeller
        Stop chopper

    Mill/transfer to drier
        Fit screen
        Open discharge valve
        Open drier feed valve
        Transfer (timer ?)

    Drying
        Heat to temperature
        Hold temperature

    Mill/discharge to hopper
        Fit screen
        Transfer to hopper
        Record hopper number.

Comments on the Process Actions

Process actions are not always physical actions executed on the process equipment. There are several process actions which are simply questions to the operator which he/she respond to; for example, clothing check asks the operator to confirm that he/she is using the correct PPE (personal protective equipment) as specified by the operating procedure for the process he/she is about to operate run. This can be done as an instruction on screen with electronic confirmation or with a paper batch sheet.

Parallel Process Actions

The list of process actions described above is not always a sequential list of actions to complete an operation. For example pre-heat FBD is started at the beginning of the stage but does not need to be completed until the batch is ready to transfer to the drier.

Process actions can be defined in some detail with start conditions, actions, alarms actions, close conditions. These actions should be as simple as possible, i.e. the smallest step in the process possible. Process actions should also be defined to make them as reusable as possible.

Common Resource

The standard allows the definition of plant as a common resource, either as shared use or exclusive use. For example, if in the SDMC the movement of materials between each process operation is by an automated materials handling system (MHS), this can be defined in the same manner. The following list illustrates how the materials handling system can be defined using a similar model structure to the main manufacturing process. The list is by no means complete but serves as an illustration of how the materials handling system can be defined. These stages become standard no matter where the movement is in the overall process; only the data associated with the move will differ (start and end locations, hopper identity, etc.).

Stage
    Operation
        Action
Move batch
    Locate batch
    Identify route
        Start point
        End point
        Route
        Route status
        Reserve route
    Set route
    Move batch
        Move trolley
        Pick up hopper

    Move batch
    Set down hopper
    De-allocate route
Clean hopper
  Locate bin

  Identify route
    Start point
    End point
    Route
    Route status
    Reserve route

  Set route
  Move bin
    Move trolley
    Pick up hopper
    Move
    Set down hopper
    De-allocate route

  Clean bin
    Valve actions
    Temperature control
    Rinse
    Reset bin status

  Remove bin from cleaning station
    Identify vacant location

  Identify route
    As above

  Set route

  Move bin
    As above

Physical model

The physical assets of batch processing facility can also be described in a hierarchical fashion (see Figure 13.12). Lower groupings are combined to form higher levels in the model. Due to the variety of systems used in industry the model also allows a grouping at one level to be incorporated into another grouping at the same level. The standard defines seven levels to the model but only the lower five are defined in detail. These five lower levels are used to describe the SDMC.

Process Cell

This is a logical grouping of equipment which includes all the equipment required to produce a batch. The process cell is not constrained to one method of control and can also include manual and administrative control.

270

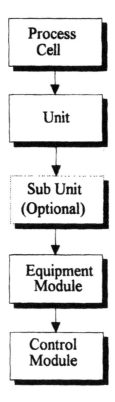

Figure 13.12 Equipment model

Unit

A unit is the equipment required to carry out one major processing activity; it comprises all the physical and control equipment required to perform that activity as an independent equipment grouping. The unit is usually centred on a major piece of processing equipment, e.g. granulator suite, compression room. There is an assumption in the standard that a unit can only operate on one batch or part of a batch at one time.

Sub Unit

This is an optional additional layer added to the ISA S88 definition of the equipment model to handle those process units that actually deal with more than one identifiable item of equipment. This also overcomes the potential problem of having two batches running on the same unit; the definition is modified to restricting one batch to a unit or sub unit if sub units have been defined for the process.

Equipment Module

The equipment module comprises control modules and if required subordinate equipment modules. The equipment module can be part of the unit, a standalone entity within the process cell, or a common resource. An equipment module can carry out a finite number of minor processing activities and contains all the physical and control equipment required to perform this activity, e.g. granulator, fluid bed drier.

Control Module

This is the basic control equipment required to carry out a single control action. This could be a temperature measurement or it could comprise a number of subordinate control modules, e.g. a temperature control loop comprising a transmitter, controller, measurement signal and setpoint for the device. It could equally be a state oriented device, e.g. an on/off switch or a header system comprising a combination of several on/off automatic block valve control modules.

Physical Model Applied to the SDMC

The physical model of the SDMC is similar to the overview diagram of the process. The process cell is defined as the SDMC; others would be LMC (liquids), IMC (inhalations) etc. The SDMC can be divided into units as indicated by the overview diagram. For this example the equipment modules and control modules have been given for the same two units of the process: granulation and bulk dispensing. These definitions are very general.

Process cell
    Units
Tablet manufacture
    Material reception
    Bulk dispensing
    Manual dispensing
    Granulation
    Blending
    Compression
    Coating
    Packing
    Dispatch
    QA.

For each piece of equipment there needs to be a series of parameters defined which not only allow the process equipment to be characterized in terms of the process functions it performs but also in terms of product restrictions. For example the majority of the equipment used for pharmaceutical manufacture is restricted with respect to which active drugs can be processed in a given process unit.

There may be a number of products containing each active, e.g. the same tablet with different dose strength, or a different type of compression die to suit a particular market's requirement.

272

Detailed Physical Model: Granulation

Unit

  Sub unit
    Equipment module
      Control module

Granulation suite
  Granulator
    Loading system
      Transfer valve
      Hopper contact
      Chute feedback.
    Solvent addition

      Load cell
      Transfer valve.

    Binder addition

      Load cell
      Transfer valve.

    Granulator

      Chopper motor
      Impeller motor
      Impeller load.

    Transfer

      Mill feedback
      Discharge valve
      Sieve size.

  Drier

    Drier
      Air flow
      Temperature
      Bowl press drop
      Filter press drop.

    Discharge system

      Dry mill
      Hopper feedback.

Detailed Physical Model: Bulk Dispensary

Unit

    Equipment module

        Subordinate equipment module

        Control module

Bulk dispensary

    Silo loading area

        Material logging

            Loading bay

            Bar code reader

            Balance.

        Transfer system

            Vacuum system

            Valve system.

  Silo

    Silo

        Feed valve

        Discharge valve

        Load cell.

  Dispensary

    Transfer system

        Diverter valve

        Vacuum system.

    Load cell

        Dispensary feed valve

        Proximity contact.

    Bulk bin

        Proximity contact.

    Lance system

        Vacuum

        Hopper feed valve.

Comments on Equipment Definition

There is a certain amount of interpretation involved in the division of equipment into modules. This is particularly true for the transfer mechanisms between two items of process equipment, e.g. the granulator and the fluid bed drier in the example above.

    The equipment definitions outlined here in combination with recipes can help to rationalize the standard operating procedures (SOPs) produced for operation of production equipment. The model gives a documented format to equipment definition thus allowing an SOP to be produced for any specific group of equipment. Recipes can then be used to define process parameter variations to cope with the minor variations in production method for different products that follow basically the same manufacturing process. The recipe can also be used to cope with any variations detected in the raw materials at any stage of the process, provided that the appropriate development work has taken place.

Following on from this, there are often duplicate process areas, each dedicated to one specific active (drug) component. This is to eliminate any possibility of cross contamination of any drug. However, the production method is the same in each production unit and could therefore be described by a common operating procedure with recipes used to define the differences is raw materials used in the batch, and the differences in process parameters used.

The packing hall is constantly undergoing change. The packing lines are generally dedicated to a product. These lines are run on a batch basis though most of the equipment can be thought of as continuous in operation with tablets going in one end and packed cartons out of the other. These lines are put together to meet the demands of individual markets; the packing line has to be able to cope with many more changes in configuration. There are typically many tens of pack variants for any one tablet product. The various packing materials should be treated exactly the same as any other raw material for the batch. The model can provide some structure and discipline to the definition of the process and rationalization of SOPs.

Batch Control Definitions

The previous sections defined the basic process and the equipment to be used; the following sections outline a structure for batch control concepts. This section describes three types of control required for batch manufacturing and the next section defines a structure for defining batch recipes.

There are three types of control required in batch manufacturing: basic control, procedural control, and co-ordination control. Of these, procedural control is the most applicable to the processes in the SDMC.

Basic Control

Basic control is the control requirements for establishing and maintaining a steady state in the process. There are several isolated examples of this in the SDMC, for example the fluid bed drier in the granulation suite where temperature and air flow have to be maintained during the drying process.

Procedural Control and Procedural Model

Procedural control is defined by means of another hierarchical structure which can be correlated with the earlier models. Procedural control directs equipment actions to take place in an ordered sequence. There are four levels to this model (see Figure 13.13). These levels are defined as follows:

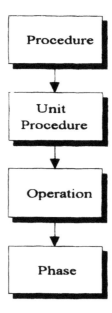

Figure 13.13   Procedural model

Procedure: The highest level in the hierarchy comprising an ordered set of unit procedures; defines the strategy to accomplish a major processing action.

Unit procedure: An ordered set of operations that allows a contiguous production sequence to take place within a unit, e.g. granulation, coating.

Operation: An ordered set of phases that defines major processing sequences that typically take the material from one state to another. Only one operation may be active in a unit at a given time.

Phase: This is the smallest element of procedural control that can accomplish a process oriented action. It may be further sub-divided into steps and transitions if necessary where a step can cause one or more actions. These can cause process changes such as: issue commands to state devices, set alarm limits, set controller constants, conduct operator authorization checks, etc. The execution of a phase may result in: commands to basic control functions, commands to other phases which may or may not be part of the same operation, data collation.

The Procedural Model in SDMC

This section outlines the procedural control structure for the SDMC. The whole process is described for the procedure and unit procedure levels and for the operation and phase definition granulation is again used to illustrate the use of these levels in the model. A brief explanation of the actions of each phase is also given. Where they have been defined the unit procedure is applied to the sub-unit in the equipment model.

Procedure

   Unit procedure

      Operation

Make tablets

Receive materials.

Bulk dispense
Manual dispense.

Granulate

    Preparation

    Load

    Dry mix

    Binder addition

    Wet mix

    Transfer (mill)

    Dry

    Discharge (mill)

    CIP.

Blend
Compress
Coat
Pack
Dispatch
QA.

From the above list the granulate unit procedure can be further expanded to the phase level, using the preparation operation as an example.

Operation
Phase	*description*
Preparation	
Area clean	Check equipment status
	Operator to confirm room status.
Clothing challenge	Operator to confirm use of correct PPE.
Ambient condition	Check temperature and humidity are within
	tolerance for the process.
Spray arm check	Ensure correct nozzle is fitted.
Binder vessel level	Ensure sufficient binder is available and
	TARE vessel.
Prime binder pipe	Manually prime the pipe.
Spray arm position	Check the spray arm is in the correct position.
Through floor feeder	Check the through floor feeder is in
	correct position.
Pre-heat FBD	Start the heating of the fluid bed drier
	to operating temperature
Load mixer	
Check bin	Confirm bulk bin batch identity and position
Open valve	Open transfer valve.

Transfer	Transfer e.g. timer
Close valve	Close transfer valve.
Dry Mix	
Start impeller	Start at correct speed
Start chopper	Start at correct speed
Start timer	
Binder addition	
Impeller speed	Change speed if required.
Chopper speed	Change speed if required.
Start pump	Add a predefined quantity over a preset time period
Run timer	Stop pump.
Wet mix	
Impeller speed	
Chopper speed	
Run timer	
Stop impeller.	
Stop chopper	At end of the operation record the quantities and actual times used
Transfer (wet mill)	
Fit screen	Ensure the correct size screen is fitted to the mill.
Open discharge valve	Open drier feed valve.
Transfer	Completion of transfer is manually determined.
Drying	
Run drier	Dry the material using pre-set inlet air temperature and volume. The product temperature is also set Dry until moisture falls to within pre-set limits.
Record results	Record the actual drying time and the final moisture content of the batch
Transfer (Dry Mill)	
Fit screen	Ensure the correct size screen is fitted to the mill.
Locate product bin	Check product bin is in place. Record bin identity.
Transfer	Completion of transfer is manually determined.
Material balance	Send bin to the weigh station on the materials handling floor and request gross weight. Perform material balance for process. If outside acceptable limits account for any variation and inform supervisor.

Co-ordination Control

This is a general function that directs and/or modifies procedural control; the types of function include:

- supervising availability of equipment
- allocating equipment to batches
- arbitrating requests for equipment
- co-ordinating common resource equipment

- selecting procedural elements for execution
- propagating modes.

This type of function has more relevance to flexible batch processes where for example a batch could be mixed in any one of a number of similar vessels. The majority of the pharmaceutical processes require dedicated equipment so this type of control and scheduling is not as relevant to pharmaceuticals as other processing industries.

The materials handling system as a common resource to the whole of the process will require some co-ordination. This function can be used for resolving conflicts that arise between batches using the MHS (Material Handling System) trolleys. Rules can be defined to allow for priorities of the various batches competing for the resource.

Recipes: ISA S88 Recipe Model, and Recipe Models

The ISA S88 standard defines a four level hierarchical model to describe the types of recipe found in a batch processing facility. These recipes are used to define a product and how that product is produced. The principle behind the different layers is that various parts of the enterprise require different levels of detail of information and may even do different things with that information. The basic recipe model is shown in Figure 13.14.

The model is designed to be as universal as possible; many batch processing plants have a selection of equipment that could be used to produce a particular batch. Therefore only the control recipe will contain detail to the level of the actual equipment that will be used to make the batch. The higher levels of the recipe will be expressed in general terms that can be modified to produce a control recipe for

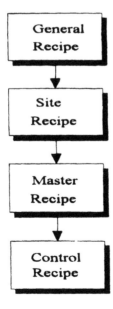

Figure 13.14   ISA S88 recipe model

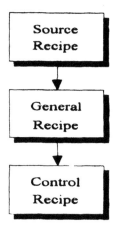

Figure 13.15 Modified recipe model

specific items of process equipment. Within pharmaceutical manufacturing there is very little flexibility in the process equipment, so the model can be modified to make it more relevant to specific requirements. Most production processes are developed within the research group and transferred across to manufacturing; consequently a new type of recipe would be required to represent the information handed over with the production method. The recipe model could be modified to suit pharmaceutical manufacturing as shown in Figure 13.15.

In this modified model the three components are:

• Source recipe:

The information received with the process from research and development and registered as part of the drug master file.

• General recipe:

This is a version of the source recipe adapted for general manufacturing. It could be used on any site required to perform the particular manufacturing process. Consequently at this level the recipe will be missing some detailed process information.

• Control recipe:

The version of the recipe used to manufacture batches at a specific location. The process equipment is fixed, therefore the control recipe will contain all the information required to manufacture the batch in the equipment available on the site for which it was written.

Recipe Contents

There is a structure defined for the contents of recipes. This structure is consistent for all levels of the recipe; the difference is the amount of detail defined at each level. In the ISA S88 standard there are five categories of recipe information identified: headers, formulas, equipment, procedures, and other information. The five categories are outlined in the following sections.

280

● *Headers*

The recipe header contains the administrative information. Typically this could contain recipe and product identification, the version of the recipe, the originator, issue data, approval date, recipe change history, reference to the high level recipe and version used to generate this recipe.

● *Formulae*

The formula information can be divided into three sections: process inputs, process parameters, and process outputs.

Process inputs describe the raw materials required to make a batch; these will also include the services required, e.g. power requirements, process services, and manpower. Inputs will include not only the product components but also the packaging materials. Quantities may be expressed as absolute quantities or as equations based on other values. Process inputs could also be used to define alternative materials where substitution is allowed.

Process parameters define information such as temperature, pressure, or timers required in the process. They could be used as setpoints, comparisons, or logic conditions.

Process outputs define all of the expected outputs from the process including expected waste products and losses in the process.

● *Equipment Requirements*

The equipment requirements contain the choice of equipment to perform specific parts of the process. The general recipe will hold this information in general terms; it will contain details of the characteristics of the equipment required and allowable materials. The control recipe will be much more detailed and will relate to a specific piece of equipment and specific lots of raw materials.

● *Recipe Procedure*

The recipe procedure defines the strategy for achieving the manufacture of a batch. The general recipe is described in non-equipment specific terms, for example temperature control.

● *Other Information*

This section contains all other relevant information not already covered. This could include hazard data, SOPs to be followed, unusual processing requests.

Recipe Model in the SDMC

Source Recipe

This recipe would contain all the data available from R&D about the process as developed. Ideally this information should be structured in the same way as the lower levels which will be used for manufacturing. This will make the generation of the lower level recipes as straightforward as possible.

General Recipe

This is a combination of the general and site recipe described by the ISA S88 standard. The general recipe is the starting level for production. It will contain the product and quantity to be made. These may be expressed in terms of a fixed batch quantity or as a standard batch size which can be modified to the actual batch size in the control recipe. By the very nature of the process this information will also fix most of the equipment to be used for production. Each section is illustrated with an example taken from the SDMC; these examples are by no means complete.

General Recipe Header

Product:	Tablet XYZ
Dose:	5 mg
Colour:	White
Product code:	XYZ123ABC
Recipe version no.:	2.0
Date:	21 June 1996
Author:	
Approved:	A. N. Other:

Revision history:

1.0 Initial version for trial
1.1 Revised after production trials
1.2 Pre-blend of dye stage added
1.3 Version used for OQ of production
2.0 Approved version following OQ/PQ

General Recipe Formula

1. Process inputs

Quantities expressed in Table 13.1 are to make a standard batch of product.

Table 13.1 Quantities required to make a standard batch of product

Item code	Description	Quantity
B124234	Component 1	4.200 kg
D576731	Component 2	9.800 kg
F703287	Component 3	39.20 kg
B786225	Component 4	3.000 kg
F235566	Component 5	156.4 kg
D536577	Component 6	0.200 kg
F685855	Component 7	4.200 kg
A545424	Component 8	51.00 L

Table 13.2 Granulation process parameters

Parameter	value	Tolerance
Binder Addition	3 minutes	± 20 Sec.
Dry Mix	5 min	± 20 Sec.
Wet Mix	3 min	± 20 Sec.
Wet Mill Screen Size	0.375 in.	
Drier Temperature	70 deg. C	± 2 Deg C
Drier Moisture Target	< 1.5%	
Dry Mix Impeller Speed	20 rpm	
Dry Mix Chopper Speed	10 rpm	
Binder Addn Imp. Speed	20 rpm	
Binder Addn Chop. Speed	1 rpm	
Wet Mix Impeller Speed	30 rpm	
Wet Mix Chopper Speed	20 rpm	
Process Yield	97–101%	

Note: The granulation process is dependent on the physical properties of the feed material. These materials can be sourced from different suppliers and may vary from batch to batch from the same supplier. Therefore the actual parameters used could be modified by the process equipment once the characteristics of the feed material are known.

2. Process parameters

As before, Granulation is used as an example (see Table 13.2). A table like this will also be produced for all the other process stages.

3. Process outputs

Item code	Description	Theoretical yield:
P88238	Tablet XYZ 5 mg White	268.0 kg

General Recipe Equipment Requirements

Again using granulation as an example:

Process Stage
   Equipment
      detail

Granulation
   Binder vessel

      Manual addition

      Heater (to 75°C)

      Mixer

      Load cell.

   Granulator

      200 kg capacity

      Variable speed impeller

Variable speed mixer

Impeller load recorder

Through floor feeder

Stainless steel.

Wet mill

Variable screen.

Drier

200 kg capacity

Operating range to 80°C

Moisture balance

Stainless steel.

Dry mill

Variable screen

General Recipe Procedure

Due to the limited variation allowed in the process this part of the recipe is the same as the process model as defined down to the Operation level.

Process Stage
    Process Operation
Raw materials handling
    Material arrives in SDMC
    QA
    Sieving
    Load bulk silo
    Transfer to holding area
Manual dispensing
    Equipment check
    Material check
    Booth dispensing
    Isolator dispensing
    Reconciliation
    Sampling
    Sieving
Bulk dispensing
    Equipment check
    Material check
    Transfer manually dispensed material
    Discharge to hopper
    Reconciliation
Binder preparation
    Water addition
    Solid addition
    Heat

Slurry preparation
    Slurry addition
    Heat
    Agitate
Granulation
    Equipment check
    Load mixer
    Dry mix
    Solvent addition (binder etc.)
    Wet mix
    Mill/transfer to drier
    Drying
    Mill/discharge to hopper
    CIP
Blending
    Transfer to blend station
    Blend
    Transfer to holding area
Compression
    Equipment check
    Material check
    Manual tablet compression
    Compression run
    Process tests
    Layering press
    Discharge
    CIP
    Coating
    Equipment check
    Material check
    Solution preparation
    Charge coater
    Coat
    Drying
    Discharge
    CIP
Packing
    Line clearance
    Material check
    Machine set up
    Material sorting
    Initial run
    Main run
    Carton
    Sampling
    Casing
    Overwrapping
    Bulk packing
    Granule packing

Control Recipe

The information given in the general recipe is taken and expanded to define everything required to produce a batch on the selected equipment.

Control Recipe Header

Product: Tablet XYZ
Manufacturing location: SDMC UK
Batch ID: XYZ2394-3
General recipe reference: Tablet XYZ Version 2.0
Dose: 5 mg
Colour: Red
Product Code: XYZ123ABC
Recipe: Version No. 2.0
Date: 21 June 1994
Author:
Approved: A. N. Other:

Revision history

1.0   Initial version for trial
1.1   Revised after production trials
1.2   Pre-blend of dye stage added
1.3   Version used for OQ of production
2.0   Approved version following OQ/PQ

Control Recipe Formula

1. Process inputs

Quantities expressed in Table 13.3 are to make a batch of product which, in this case, is exactly half the standard batch amount.

2. Process parameters

As before Granulation is used as an example (see Table 13.4). A table like this will also be produced for all the other process stages.

Table 13.3   Quantities required to make a specific batch of product

Item code	Description	Quantity	Lot identity
B124232	Component 1	2.1 kg	WJ4521
D576731	Component 2	4.900 kg	WJ5483
F703287	Component 3	19.60 kg	NJ4979
B786225	Component 4	1.500 kg	WJ7563
F235566	Component 5	78.20 kg	NJ4958
D536577	Component 6	0.100 kg	WJ3710
F685855	Component 7	2.100 kg	NJ3847
A545424	Component 8	25.50 L	HJ3649

Table 13.4 Granulation process parameters for a specific batch

Parameter	Value	Tolerance
Binder addition time	3 min	±20 sec
Dry mix time	5 min	±20 sec
Wet mix time	3 min	±20 sec
Wet mill screen size	0.375 in	
Drier temperature	70°C	±2°C
Drier moisture target	<1.5%	
Dry mix impeller speed	20 rpm	
Dry mix chopper speed	10 rpm	
Binder addn imp. speed	20 rpm	
Binder addn. chop. speed	10 rpm	
Wet mix impeller speed	30 rpm	
Wet mix chopper speed	20 rpm	
Process yield	97–101%	

Note: The granulation process is dependent on the physical properties of the feed material. These materials can be sourced from different suppliers and may vary from batch to batch from the same supplier. Therefore the actual parameters used will change.

3. Process outputs

Item code   Description: Theoretical yield:
P88238   Tablet XYZ 5 mg White   134.0 kg

Note: In pharmaceutical processes there are two equally important products of the manufacturing process; one is the product itself and the second is the supporting documentation generated during the manufacture. ISA S88 does not make this distinction in the models and functionality it describes. However this can be readily accommodated within the functions it does define by including documentation requirements in the recipe. Completed documentation from one process stage can be included in the list of process inputs for the next process stage and completed documentation for the stage included in the list of process outputs. The inclusion of documentation in the recipe makes no assumptions about whether the batch records are in electronic or manual (paper) form.

Control Recipe Equipment Requirements

Again using granulation as an example:

Process Stage
 Equipment
  Detail
Granulation
 Binder vessel
  Manual addition
  Heater (to 75°C)
  Mixer
  Load cell.

Granulator

200 kg capacity

2 speed mixer

Impeller load recorder

Through floor feeder

Stainless steel.

Wet mill

0.375 in screen.

Drier

200 kg capacity

Operating range to 80°C

Moisture balance

Stainless steel.

Dry mill

0.375 in screen.

Control Recipe Procedure

Process Stage
   Process Operation
Raw materials handling
   Material arrives in SDMC
   QA
   Sieving
   Load silo
   Transfer to holding area
Manual dispensing
   Equipment check
   Material check
   Booth dispensing
   Isolator dispensing
   Reconciliation
   Sampling
 · Sieving
Bulk dispensing
   Equipment check
   Material check
   Bulk dispense
   Transfer manually dispensed material
   Discharge to MHS bin
   Reconciliation
Binder preparation
   Water addition
   Solid addition
   Heat

Slurry preparation
Slurry addition
Heat
Agitate
Granulation
    Equipment check
    Load mixer
    Dry mix
    Solvent addition (binder etc.)
    Wet mix
    Mill/transfer to drier
    Drying
    Mill/discharge to hopper
    CIP
Blending
    Transfer to blend station
    Blend
    Transfer to holding area
Compression
    Equipment check
    Material check
    Manual tablet compression
    Compression run
    Process tests
    Layering press
    Discharge
    CIP
Coating
    Equipment check
    Material check
    Solution preparation
    Charge coater
    Coat
    Drying
    Discharge
    CIP
Packing
    Line clearance
    Material check
    Machine set up
    Material sorting
    Initial run
    Main run
    Carton
    Sampling
    Casing
    Overwrapping
    Bulk packing
    Granule packing

1. Hazards:
   This drug is very potent and is absorbed through the skin, by inhalation and if swallowed. The drug has these effects in limited and repeated exposure.
2. Protective clothing:
   Follow SOP 123ABC. An airflow helmet MUST be worn through dispensing, bin loading, granulation, drying and milling. An airflow helmet MUST be worn in compression when in contact with the granule and when taking samples from the keg. Plastic or rubber gloves must be worn at all stages of the process.
3. Spillage:
   Using the clothing specified above collect the spillage with a vacuum cleaner and put into a labelled container for disposal. Wash spillage site with water.

Conclusions

The examples given in the previous sections clearly illustrate how the ISA S88 standard can easily be applied to a secondary pharmaceutical manufacturing plant. The standard gives a formal structure to the process as it already exists. This will make the definition and design of interfaces between the plant and higher level supervisory and planning systems much easier. The definition of data structures and recipes will be more consistent if each part of the manufacturing process is defined using the same terminology and hierarchy.

It is not possible to quantify the benefits of using the ISA S88 standard from this exercise; that can only be done from the results of the implementation of a real system. However, once a process is defined using a fixed set of terminology there is scope for reducing confusion and errors when designing and implementing control systems on existing equipment.

In reality, where there are specific software systems selected for manufacturing execution systems, there will be a set of terminology already defined. Similarly many existing plants will have their own terminology that has evolved with the plant; this may be several sets of terminology that have grown up with the different groups working and supporting the facility, e.g. operating staff, engineers and development staff all work with their own terminology. ISA S88 gives a good starting point for defining common terminology. Even if the ISA S88 definitions are not followed exactly the discipline associated with the use of these models is still of considerable benefit.

The recipe structure gives a good basis for defining where each level recipe is defined and therefore which systems are responsible for determining that information.

Terminology has been defined with a standard set of definitions. However, in practice this becomes difficult to adopt rigorously because software packages will set their own terminology rules and existing work culture will also give rise to common local terminology.

The discipline and structure given by the standard can easily be adapted to local terminology. It is the discipline rather than the words used that is the benefit to be gained from the standard.

References

Guide to Good Manufacturing Practice, 1983 (*The Orange Guide*), Department of Health and Social Security.

KANIG, J. L. and RUDIC, E. M., The basics of robotic systems, *Pharmaceutical Technology*, June 1986.

LUMSDEN, B., Industrial Powder Technology Conference, London, 24–25 November 1982.

MORLEY, R. E., *The Role of Control in Automation Trends and Perspectives*, Jan–Feb, 1984.

ISA S88.01–1995 Standard (approved February 28, 1995) Batch Control Part 1: Models and Terminology ISBN 1-55617-562-0.

14

Packaging Systems

'I see you're admiring my little box', the Knight said in a friendly tone.
'It's my own invention – to keep clothes and sandwiches in. You see I carry it
upside-down, so that the rain can't get in.'

The pharmaceutical industry is conservative in its approach to formulation, use of
new equipment, and new materials, and this is largely due to the time-consuming
process of registering products and their packs. However, it is claimed that even in
the ethical market an innovative pack can increase sales. However, total pack safety,
i.e. tamper proof, child resistant containers, sterility, and the development of original
pack dispensing has meant large investments in new types of equipment and technologies to operate them. Some examples are:

- bar codes tracking
- radio frequency tagging
- machine vision
- laser printers
- robotics
- automated guided vehicles (AGVs)
- linear transfer vehicles
- process control and monitoring systems
- material control and reporting systems
- batch documentation.

This is not an exhaustive list, but it gives some indication of the new technologies
that have to be considered.

It is not proposed to deal in any great detail with specific equipment for the
general operation of packaging dosage forms, but to discuss some ideas that are
current for moving from mechanization into automation.

The prime objective in all packaging operations must be to accelerate improvements in operating productivity while at the same time maintaining absolute assurance of quality. The procedures that can be considered for automation include line
monitoring, the feeding of components, bottle filling, inspection of labels, removal of
nonconforming packs, and the automation of outer packaging.

Generally, in packaging operations, there are a number of tedious functions such
as checking packs for missing caps or labels, badly printed overprinting, or some

other physical fault concerned with inspection and quality assurance/control (QA/QC). The industry recognizes that to improve quality this task is best undertaken by a machine vision system.

In addition to automating manipulative activities, a process monitoring computer can perform a number of supervisory functions such as checking for low hopper levels, low level of labels, and faults such as blocked feed lines for bottles or components.

The use of a built-in diagnostic computer system can identify faults due to components failing or not meeting the required specification.

The success rate of humans in inspection lies around the 85% level; machines can achieve 100% accuracy. Moreover, a machine will operate around the clock with high reliability. But the use of machine vision systems present problems when scanning labels, and some redesign of printing may be required, for example, the machine vision system may have difficulty in scanning a 3 and an 8. A laser printer can overcome this problem. Therefore, the use of robots to handle some of the routine tasks on packaging lines coupled with machine vision can increase productivity and QA by a considerable degree.

For these reasons, companies like Merck, Sharp, and Dohme (MSD), Roche, Ciba-Geigy, and SmithKline & French (SKF) now SmithKline Beecham, in Europe are using these systems to increase output on a more reliable basis. The motivation can be summarized for reducing the labour content of processing and packaging operations:

- Increased yields.
- Better reliability.
- High cost of labour and its unpredictability.
- Elimination of boring tasks.

On a conventional packaging line using a slat counter the prime requirement is the accurate filling of tablets into bottles. Traditionally, the line comprises a number of complex units linked by a conveyor and accumulators for moving the product along the line and pacing the flow. Here there are a number of options available for automation. The primary machines will fill bottles with the dosage form, pack the bottles, and bundle packages. These machines and their auxiliary equipment are usually electromechanically coordinated to ensure proper assembly and packaging of the product.

Sections of this type of line targeted for automation should include bottle orientation, capping, labelling, and collation (bundling). Generally on a non-automated line an operator is responsible for checking bottles of product passing the labelling station, probably at a rate approaching 100 per minute, for missing caps, missing or incorrect labelling, and missing or illegibly overprinted information.

In automating this operation an extremely tedious job could be eliminated and improved product security obtained (quality assurance).

At the end of the line, another function which can be considered for automation is the transfer of finished packs from the line to a caged pallet. This would appear to be an ideal application for a robot, but it may be difficult to find a model which could be modified sufficiently to give the flexibility required for its operation, and some form of transfer conveyor may provide a better option. Most robots are designed for use in other industries and consequently do not have the reach or speed to be able to handle packages coming off a pharmaceutical packaging line and be economic to instal.

Automation of this type of packaging will also require a system for monitoring the operation that will cover a number of supervisory functions, for example, checking for low hopper levels, fallen bottles, low label supplies and its routine functions. Wherever possible, machines should be equipped with local logic, i.e. their own microprocessors instead of a large central processing unit to perform product surveillance and rejection. In addition it is necessary in the case of a back-up of product flow to supply product to the upstream machine if this is interrupted, as an alternative to turning off the machine. This avoids problems of having to automatically restart the machines where contactors have been tripped. It is also necessary to design the system with reject gates operating in a fail safe mode, that is all products would be rejected unless an accept signal switches them away from the reject line.

In a line similar to this at least 70 sensors will be needed to indicate low levels in hoppers, position of safety guards, supply of process utilities such as air pressure, and the status of the motors. All these sensors should be linked to a programmable logic controller (PLC) which scans each sensor several times each second and compares the readings with a preset value. When a discrepancy occurs, a warning is displayed on the screen and the event is recorded by the microprocessor connected to the PLC. In addition to displaying an up-to-date report on the status of the line, the microcomputer provides other functions: it prints a record of rejections and machine stoppages at the end of the batch and archives the information for historical purposes. Secondly, the set-up procedure requires that an operator at the keyboard enters the identity number of each component to be used – bottles, caps, labels, etc. The microcomputer compares these numbers with preset values to ensure that the correct components have been drawn from the stores and are in use on the line.

One of the most tedious jobs on the packaging line is the control and scanning of the labels that are being used. It is possible to purchase a software package that checks the labels by bar code and watches for missing labels and the absence of overprinted batch numbers as well as providing a system for appropriate rejection. One of the difficulties here is the fact that the legibility of the overprinted information can be difficult to scan. One solution to this problem is to use a laser printer. This printer is essentially a stencilling device that uses a mask of the batch number and a light beam. Proper firing of the laser is confirmed by a device in the printer that checks the voltage built-up in the unit. While this is an expensive option it does eliminate the need for inspection of the overprinted codes, since if the laser fires, then it is known that the codes are legible (the system must be validated), but if it does not fire the PLC issues an alarm and ensures that package is rejected.

An alternative system has been developed which uses computerized machine vision. This comprises linking a television camera and a computer to read the identification number printed on the label, and compare it with a preset value. This can also be used to read the on-line overprinted information comprising the batch number, manufacturing date and expiration date. However, one of the difficulties here was the task of character recognition (e.g. 3 and 8), and this proves more difficult in practice than can be immediately recognized; for instance, it is sometimes difficult to read or distinguish between a 5 and a 6. The system must not be so inflexible as to reject characters which, although not perfect, are legible. Better discrimination is always possible by using more complex algorithms, but this takes more computer time.

Included in the automation programme should be the requirement to generate batch documentation. The operators working on the line, and the quality assurance inspectors, all need to input information into the system at various times during the packaging operation. At completion a printout of the batch records could then be produced.

In addition to the conversion of a manually operated slat line packaging system to an automated line, there are other major products that require similar treatment:

- Liquid oral dosages
- Powders
- Capsules
- Transdermals
- Aerosols
- Ointments
- Suppositories
- Medical devices such as pacemakers
- Large volume parenterals (LVPs)
- Small volume parenterals (SVPs)

In the case of tablets and capsules, blister packaging has become the preferred system, and companies such as Uhlmann have developed lines to provide a complete automated system with built-in diagnostic fault finding and batch reporting systems. For a blister the remaining problems are how to handle the packaging components, for example:

- rolls of aluminium foil and polymers such as PVC and PVDC
- cartons
- leaflets
- outer carton and overwrappers.

The options are:

- overhead gravity feed system
- use of fork lift trucks (FLTs) or AGVs
- consolidation of materials prior to start of operation and reconciliation at completion
- clean room operations
- label and printed material control
- inventory philosophy.

However, the degree to which automation has been implemented in packaging operations has been limited to a reduction in numbers of operations in the equipment made possible by greater control sophistication on the units and transfer conveyors for feeding automated carton glueing machines before transfer to automated warehouse systems.

15

Validation

*I heard him then, for I had just
Completed my design
To keep the Menai bridge from rust
By boiling it in wine.*

This chapter describes the activities that are necessary to achieve the validation of typical manufacturing facilities to FDA and MCA requirements. The activities for the validation programme can be summarized as follows:

- Development of master validation plan for the facility.
- Design review of the facility, utilities, and process equipment.
- Protocol development for the facility, utilities, systems, and processes.
- SOP development.
- Validation of analytical test procedures.
- Calibration of instruments and equipment.
- Training in good manufacturing practices (GMPs), standard operating procedures (SOPs), and validation.
- Establishment of equipment history file.
- Administration, direction, scheduling, guidance, and execution of the physical validation.
- Review, evaluation, certification of the validation data.

Schedule of Activities

The general schedule of activities and an approximate time framework for a typical validation programme are depicted in Figure 15.1. It should be recognized that Figure 15.1 is a preliminary schedule which evolves over time. The activities listed are strictly dependent upon the construction schedule, specific nature of the facility, and particular validation requirements, the constraints of which must be incorporated as the information becomes available.

There is no single designers' handbook related to the manufacture of pharmaceutical or dosage forms. However, there is a collection of guides, standards, and knowhow that must be embodied in the design, construction, commissioning and

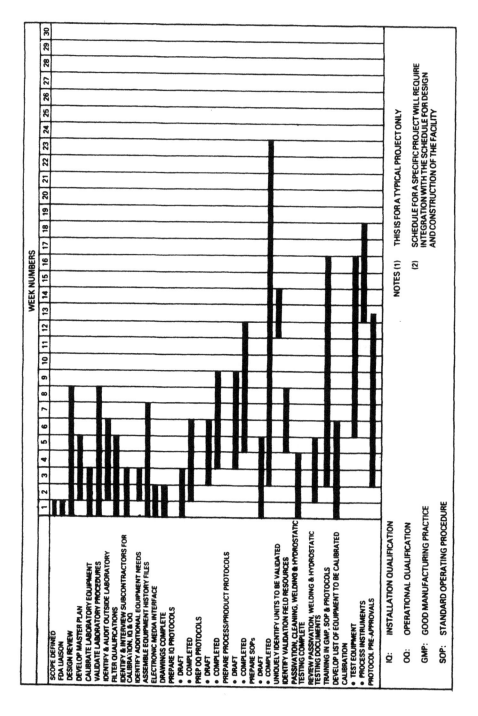

Figure 15.1 Schedule for typical validation project

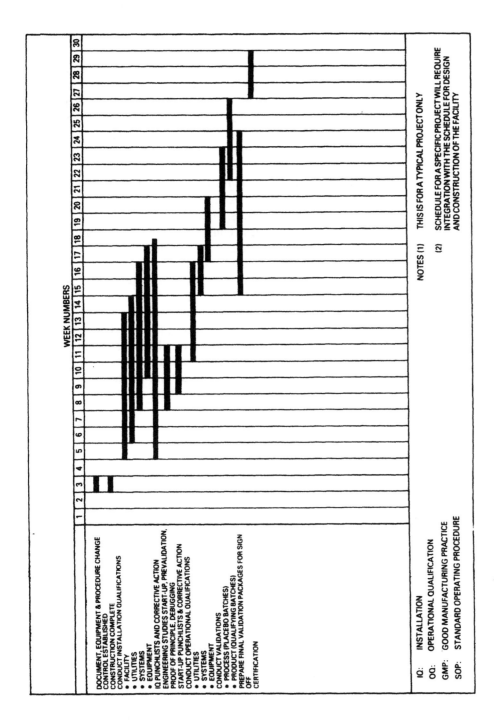

Figure 15.1 (continued)

operation of a facility. In the UK the prime documents available to the designer are *The Guide to Good Manufacturing Practice* 1983 (*The Orange Guide*), British Standard 5295, and US Federal Standard 209E, which is particularly applicable for clean rooms.

A typical example of the application of these standards is the increasing need within the pharmaceutical industry to manipulate toxic or hazardous substances. In these cases the designer of the facility has to overcome the necessary compromise between effective hazard containment and secure product protection, and therefore the designer and facility user have to derive an effective solution that meets these conflicting demands of containment and contamination. All of this has to be addressed in the validation programme.

Initially, the effort will focus on a design review and document development which commences with the preparation of a master plan, installation qualification (IQ), operation qualification (OQ), and process validation (PV) protocols for the facility and all utilities. Protocol development continues then with the preparation of documents for the various systems and processes at the facility. This same order of prioritization is repeated during the execution stage of the physical validation to ensure that entities undergoing validation are supported by previously validated utilities.

Before individual 'qualifications' are considered some up-front activities must be addressed before embarking on a prospective validation exercise. They are not all validation related, but each will play its role in the overall success of the project. These activities are not so applicable to retrospective validation exercises but some may apply if shortcomings are revealed and remedial actions are necessary. These are:

- Generation of user requirement specifications.
- Identification of 'system' boundaries.
- Identification of resource requirements.
- Specification of documentation to be provided by 'system' vendors.
- Initial assessment of 'system' vendors.
- Initial GMP review of 'system' designs.

Some, or all, of these may be treated quite informally and may fall outside the bounds of the Validation Master Plan. That is a judgement. The important issue is that they are addressed.

User Requirement Specifications

User requirement specifications (URS) should be prepared, by the 'system' to be validated, in terms of the final process requirement. A URS should typically provide specific, but non-detailed information relating to, for example, quantity, quality, compatibility, performance, environment, finishes, etc. in terms of:

- materials of construction
- cleanability requirements
- maintenance requirements
- performance criteria

- critical parameters
- operating ranges of critical parameters
- essential design criteria
- requirements of computer systems
- training and documentation requirements,

and should make reference to all relevant in-house standards and regulatory documents.

Master Plan

The master plan serves a dual purpose as:

1. A document which may be presented to regulatory bodies to convey the level of understanding of company responsibilities concerning the validation programme, along with plans to discharge that responsibility.
2. A guide to those administering and performing validation activities.

The master plan will address and include, but not necessarily be limited to, the following topics:

- Approvals
- Introduction
- Scope
- Glossary of Terms
- Preliminary drawings/facility design
- Raw material qualification
- Process descriptions
- Rooms and room classifications
- Description of utilities
- Description of process equipment
- Automated systems
- Equipment history files
- Construction documentation
- Description of required protocols
- Lists of standard operating procedures (SOPs)
- Required document matrices
- Validation schedules/construction schedule
- Integrated schedule
- Protocol outlines/summaries
- Environmental monitoring
- Analytical testing procedures
- Calibration programme

Pharmaceutical Production Facilities

- Training programme
- Preventive maintenance programme
- Change control programme
- Document control programme
- Manpower requirements
- Key personnel
- Protocol examples
- SOP examples.

Design Review

As part of the overall design review it is necessary to ensure that the total facility as designed is validatable.

Protocol Development

Typical protocols required for the validation effort are shown in Table 15.1, which shows a total of 15 installation qualifications (IQ), 14 operation qualifications (OQ), and 6 process/product validations (PV) protocols that are needed for the various utilities, systems, and equipment.

The protocols identified in this proposal must be tailored specifically for the facility.

Table 15.1 Protocols for facilities, systems, and processes

	IQ	OQ	PV
Facility	X		
Utilities			
Emergency electrical generator	X	X	
LTC chiller	X	X	
Deionized water	X	X	X
HVAC	X	X	
Nitrogen	X	X	X
Process steam	X	X	
Chilled water	X	X	
Vacuum cleaning system	X	X	
Vacuum drying system	X	X	
Systems			
Computerized stock control	X	X	
Processes			
A1	X	X	X
A2	X	X	X
A3	X	X	X
A4	X	X	X

Particular attention must be paid to the computerized systems' documentation requirements. Protocols for a 'stock control' QA status and material handling system must be developed as separate entities as shown in Table 15.1. Validation of a computerized process control system can be handled as part of the IQ, OQ, and PV protocols for three intermediary products and the final product.

SOP Development

Standard operating procedures (SOPs) need to be in place and some development is usually required to ensure compatability with the master plan.

Analytical Support

All analytical support of the validation effort is the responsibility of the internal laboratory or a qualified outside laboratory.

All laboratory methods need to be validated before the physical validation. Development of methodology for validating analytical procedures must occur early in the programme.

Calibration

The calibration of all instruments, gauges, and equipment must be performed before the onset of the physical validation by the company or a qualified sub-contractor.

Training

The development of a viable training programme must be in place before validation takes place. The company has the responsibility for training personnel in SOPs and GMPs before validation. Training in validation should be handled as part of the protocol review process for the companies' personnel, and would be recorded as such.

Equipment History File

The establishment of equipment history files plays a central role in the timing and orderly execution of validation activities. These records are instrumental in providing the validation team with the documentation required to complete the validation protocols and certify utilities, systems, and processes.

All the equipment history files for all items subject to validation must be assembled. These files consist of specifications, purchase orders, invoices, receipts, receiving records, certificates, performance curves, manuals, drawings, vendor product information, and test results, and any other relevant documentation. The assembly of the equipment history files is an activity that continues through the construction, start-up, and validation phases of the project. Maintenance of these files continues beyond validation.

Physical Validation

Physical validation can commence when the following activities have been completed:

- The design review has been completed.
- All SOPs relating to the validation effort have been written.
- All appropriate SOP training has been conducted and documented.
- Document, procedure, and equipment change control is in place.
- Equipment history files including purchase orders, drawings, etc., are complete.
- All 'as built' drawings (layouts, piping, and instrumentation drawings, isometrics, etc.) are complete.
- All passivation, welding, and hydrostatic testing has been completed and documented.
- All instruments, gauges, test equipment, etc., have been recently calibrated.
- The protocols have been completed and accepted.
- The descriptive information (equipment identification, SOP numbers, etc.) required by the protocol has been entered and signed off as approved.

The physical validation begins with the execution of the installation qualifications for the facility and for those utilities such as HVAC, which supports equipment and other utilities. The implementation of the operation qualifications for these utilities begins as soon as the IQ has concluded. This same pattern will follow for the remaining entities in the order: other utilities, systems, and processes as enumerated in Table 15.1.

All field work assumes the validatability of the facility, utilities, systems, and processes and the use of the standard industry criteria of three consecutive successful trials.

The FDA defines validation as 'Establishing documented evidence which provides a high degree of assurance that a specific process, system, or entity, will consistently produce a product meeting its predetermined specifications and quality attributes' (well structured, well documented, common sense. Ken Chapman, Pfizer Inc.). Validation has also been defined as the activity performed to demonstrate that a given utility, system, process, or piece of equipment does what it purports to do. The primary means of accomplishing this end is the scientific study designed to specifically permit the determination as to whether the entity under scrutiny in fact:

- Meets or exceeds the specifications of its design.
- Is properly built, shipped, received, stored, installed, operated and maintained.
- Is suitable for its intended application.
- Is in accordance with principles established and generally accepted by the scientific community.
- Conforms to basic cGMP design criteria.

- Will satisfy the concerns of regulatory bodies.
- Is capable of consistently producing a product that is fit for use.
- Will meet the goals established for productivity, safety, and quality.

This scientific study is generally detailed in a validation protocol. A well designed validation programme properly supported by senior management will accrue considerable benefit to its sponsor. Not only will regulatory obligations be fulfilled but also processes will be optimized, productivity improved, and downtime reduced. In short, a validation programme with a sound scientific base and proper experimental design is simply good business if taken seriously and executed conscientiously.

The purpose of the validation master plan is to convey to the FDA and other regulatory bodies, understanding of a company's responsibilities concerning the validation and certification of the facility along with plans to discharge that responsibility. It will also serve as a guide to those administering and performing validation activities and as a road map to successful project completion. The master plan may be used as an instrument to define areas of responsibility and accountability to validation team members, and it contains all the programmes necessary to certify the validation of the facility system and processes.

By integrating the pharmaceutical and engineering disciplines, the validation master plan describes the facility and outlines a set of activities tailored to that facility that provides a cost effective validation programme that not only minimizes regulatory exposure but is delivered within the prescribed time constraints.

The life cycle concept of validation is a recurring theme throughout the master plan. Paramount is the establishment of the infrastructure to ensure that the facility is supported by sufficient documentation throughout its conceptual and functional lifetime. This lifetime includes project inception, design, engineering, construction through testing, certification, maintenance, revalidation, and change control.

Scope

The validation master plan is prepared for a facility, which when completed, validated, certified, and approved will be used to produce pharmaceutical products. These products can be a bulk pharmaceutical product or various dosage units.

Generally, the facilities to be validated consist of rooms which are to be dedicated to the formulation and manipulation by chemical reaction or unit operation such as mixing, blending, compaction, filling, and packaging of products. The environments of these facilities can be designated as Classified: controlled; Classified: critical; and Unclassified. Certain requirements such as Class 100 and Class 10000 for specialized clean rooms and sterile suites may be required. The operational area is typically distributed among the following: a warehouse, a processing area for unit operations, filling and packaging areas, analytical laboratories for quality control operations, offices for quality assurance, offices, and amenities.

For the purposes of planning, validation of a site can be organized into facilities, utilities, systems, process equipment, and processes.

Among the most relevant of the regulatory issues from the *Code of Federal Regulations*, Volume 21, that should be considered in the assembly of the master plan are the following:

Part 58   Good Laboratory Practice for Non-Clinical Laboratory Studies

Part 210 Current Good Manufacturing Practice in Manufacturing, Processing, Packaging, or Holding of Drugs; General

Part 211 Current Good Manufacturing Practice For Finished Pharmaceuticals

Part 212 Current Good Manufacturing Practice in the Manufacture, Process, Packaging, or Holding of Large Volume Parenterals For Human Use (Proposed)

Part 600 Biological Products: General

Part 606 Current Good Manufacturing Practice For Blood and Blood Components

Part 610 General Biological Products Standards

Part 820 Good Manufacturing Practice For Medical Devices: General

It is also necessary to take into consideration various guidelines and manuals published by the FDA, NIH (National Institute of Health) guidelines, and OSHA organization.

The master plan should start with a table of contents followed by an approval page and a glossary that ensures all members of the validation project team understand the terminology. A typical table of contents was discussed in the Master Plan section, earlier in this chapter.

The approval page may require the participation of each or some of the following functional areas:

	Name	Signature	Date
Manufacturing			
Engineering			
Quality assurance			
R&D			
Safety			
Site manager			
Regulatory affairs			
Validation manager			

The completion of this page indicates review of the contents by the relevant disciplines and approval by responsible individuals.

Computer Validation

The validation of computers has been given a particular focus by the United States FDA.

Three documents have been published for agency and industry guidance. In February 1983, the agency published the *Guide to Inspection of Computerized Systems in Drug Processing*; in April 1987, the *Technical Reference in Software Development Activities* was published; on 16 April 1987, the agency published *Compliance Policy Guide 7132* in *Computerized Drug Processing: Source Codes for Process Control Application Programmes*.

In the inspection guide, attention is called to both hardware and software, some key points being the quality of the location of the hardware unit as to extremes of environment, distances between CPU and peripheral devices, and proximity of input devices to the process being controlled; quality of signal conversion, e.g., a signal converter may be sending inappropriate signals to a CPU; the need to systemati-

Responsible for
Specification

Responsible
for testing

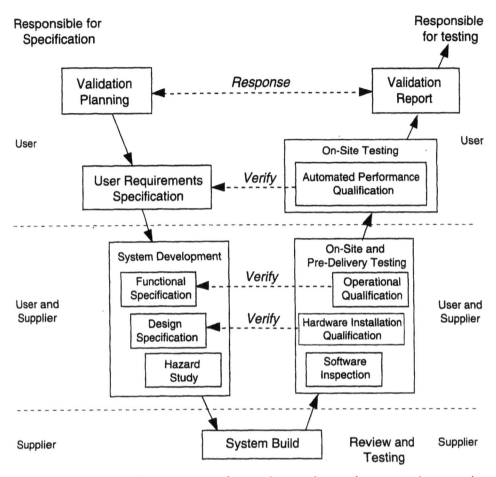

Figure 15.2 The relationship between specification, design and testing for automated systems: the Life-Cycle V model

cally calibrate and check for accuracy of I/O devices; the appropriateness and compatibility within the distributed system of command overrides, e.g. can an override in one computer controlled process inadvertently alter the cycle of another process within the distributed system? Maintenance procedures are another matter which interests the agency during an inspection. Other matters of concern are methods by which unauthorized programme changes are prevented, as inadvertent erasures, as well as methods of physical security.

Hardware validation should include verification that the programme matches the assigned operational function. For example, the recording of multiple lot numbers of each component may not be within the programme, thus second or third lot numbers of one component may not be recorded. The hardware validation should also include worse case conditions, e.g. the maximum number of alphanumeric code spaces should be long enough to accommodate the longest lot numbering system to be encountered. Software validations must be thoroughly documented – they should include the testing protocol, results, and persons responsible for reviewing and approving the validation. The FDA regards source code, i.e. the human readable form of the programme written in its original programming language, and its supporting documentation for application programmes used in any drug process control to be part of the master production and control records within the meaning of 21CFR parts 210, 3211 (*Current Good Manufacturing Practice Regulations*).

As part of all validation efforts, conditions for revalidations are a requirement.

In recent years there has been a considerable development of means of validating automated systems. The Pharmaceutical Industry Computer Systems Validation Forum (PICSVF) introduced its Good Automated Manufacturing Practice (GAMP) document. This provides suppliers with information on how the industry wants its control system designed, tested and validated. A typical V model is shown in Figure 15.2 which illustrates the basic framework.

Copies of GAMP are available from the International Society of Pharmaceutical Engineers (ISPE) in Holland.

Cleaning Validation

This is a topic that has been the subject of many papers in recent years. References (Mendenhall, 1989, Jenkins *et al.*, 1994; Fourman *et al.*, 1993) provided an overview and the FDA paper revised in 1993, developed a systematic approach to proving the effectiveness for all cleaning procedures. Cleaning procedures and their effectiveness has been and still is, the subject of much debate within the industry. It is of particular concern to all manufacturing operations, where a large number of different products are made. As the detection level improves with the development of more sensitive analytical procedures, the focus is directed at the sampling techniques. A question that needs to be asked is how clean is clean? (Hwant *et al.*, 1995, 1997) The reader is recommended a paper on the hands-on approach by Malcolm Olver of Glaxo Wellcome, presented at the Process and Computer Validation Meeting in Manchester (Olver, 1997).

Glossary

This is not meant to be a complete list, and will vary depending on individuals and their companies.

Acceptance criteria
The product specifications and acceptance/rejection criteria, such as acceptable quality level, and unacceptable quality level, with an associated sampling plan, that are necessary for making a decision to accept or reject a list or batch (or any other convenient sub-groups of manufactured units).

Action levels
Levels or ranges which, when deviated from, signal a potential drift from normal operating conditions; these ranges are not perceived as being detrimental to end-product quality.

Audit
An audit is a formal review of a product, manufacturing process, equipment, facilities, or systems for conformance with regulations and quality standards.

Bulk drug substance
Any substance that is represented for use in a drug and that, when used in the manufacturing, processing, or packaging of a drug, becomes an active ingredient or a finished dosage form of the drug, but the term does not include intermediates used in the synthesis of such substances.

Bulk pharmaceutical chemical
Any substance which is intended for use as a component in a 'drug product', or a substance which is repackaged or relabelled for drug use. Such chemicals are made by chemical synthesis, by processes using fermentation, or by recovery from natural (animal, mineral, or plant) materials.

Calibration
Comparison of a measurement standard or instrument of known accuracy with another standard or instrument to detect, correlate, report, or eliminate by adjustment any variation in the accuracy of the item being compared.

Certification
Documented statement by qualified authorities that a validation event has been done appropriately and that the results are acceptable. Certification is also used to denote the acceptance of the entire manufacturing facility as validated.

Change control
A formal monitoring system by which qualified representatives of appropriate disciplines review proposed or actual changes that might affect validated status and take preventive or corrective action to ensure that the system retains its validated state of control.

Concurrent validation
Establishing documented evidence that the process, which is being implemented, can consistently produce a product meeting its predetermined specifications and quality attributes. This phase of validation activities typically necessitates careful monitoring/recording of the process parameters and extensive sampling/testing of the in-process and finished product during the initial implementation of the process.

Installation
The documented evaluation of the construction or assembly of a piece of equipment, process, or system to assure that construction or assembly agrees with the approved specification, applicable codes and regulations, and good engineering practices. The conclusion of the evaluation should decidedly state the equipment, process or system was or was not constructed in conformance with the specifications.

Critical areas
Areas where sterilized product or container/closures are exposed to the environment.

Critical process variables
Those process variables that are deemed important to the quality of the product being produced.

Critical surfaces
Surfaces which come into contact with sterilized product or containers/closures.

The environment of a facility which is considered to be a critical area is usually limited to 300 000 particles > 0.5 μm per cubic metre. The surfaces in which the items are processed or exposed are limited to 3000 particles of > 0.5 μm per cubic metre.

D-value
The time at a given temperature needed to reduce the number of microorganisms by 90%.

Design review
A 'design review' is performed by a group of specialists (such as an architect, HVAC engineer, process engineer, validation specialist, and others) to review engineering documents to assure that the engineering design complies with the cGMPs for the facility, i.e. medical device, biologic, drug product, bulk pharmaceutical chemical, or components. The thoroughness of the design review depends upon whether the engineering project is a feasibility study, a conceptual design, preliminary engineering, or detailed engineering. Minutes of all meetings for design review will be sent to team members and the client to show the status of the design to cGMPs.

Drug
Articles recognized in the official USP, or NF, or any supplement to them; articles intended for use in the diagnosis, cure, mitigation, or prevention of disease in man or other animals; articles (other than food) intended to affect the structure or any function of the body of man or other animals; articles intended for use as a component of any articles specified above; but does not include devices or their components, parts, or accessories.

Drug product
A finished dosage form, for example tablet, capsule, solution, etc., that contains an active drug ingredient generally, but not necessarily, in association with inactive ingredients. The term also includes a finished dosage form that does not contain an active ingredient but is intended to be used as a placebo.

Dynamic attributes
Dynamic attributes are classified into functional, operational, and quality attributes.

Edge of failure
A control or operating parameter value that, if exceeded, may have adverse effects on the state of control of the process and/or on the quality of the product.

Facilities
Facilities are areas, rooms, spaces, such as receiving/shipping, quarantine, rejected materials, approved materials warehouse, staging areas, process areas, etc.

Functional attributes
Functional attributes are such criteria as controls, instruments, interlocks, indicators, monitors, etc., operating properly, pointing in the correct direction, valves allow flow in correct sequence, etc.

Good manufacturing practices (GMPs)
The minimum requirements by law for the manufacture, processing, packaging, holding, or distribution of a material as established in Title 21 of the *Code of Federal Regulations*. Examples include: Part 211 for finished pharmaceuticals; Part 606 for blood and blood components; Part 820 for medical devices.

Installation qualification protocol
An installation qualification protocol (IQ) contains the documented plans and details of procedures which are intended to verify specific static attributes of a facility, utility/system, or process equipment. Installation qualification when executed is also a documented verification that all key aspects of the installation adhere to the approved design intentions and that the manufacturer's recommendations are suitably considered.

Intermediate (drug/chemical)
Any substance, whether isolated or not isolated, which is produced by chemical, physical, or biological action, at some stage in the production of a bulk pharmaceutical chemical, and subsequently is used at another stage in the production of that chemical.

Life cycle
The time frame from early stages of development until commercial use of the product or process is discontinued.

Master plan
The purpose of a master plan is to give the FDA a document which demonstrates a company's responsibility and intent to comply with cGMPs and itemizes the elements which will be completed between the design of engineering and plant start-up. A typical master plan may contain, but is not limited to, the following elements: Approvals, Introduction, Scope, Glossary of Terms, Preliminary Drawings/Facility Design, Process Description, List of Utilities, Process Equipment List, List of Protocols, List of SOPs, Equipment Matrices, Validation Schedule, Protocol Summaries, Recommended Tests, Calibration, Training, Manpower Estimate, Key Personnel (Organization Chart and Résumés), Protocol Examples, SOP Examples.

Operational attributes
Operational attributes are such criteria as a utility/system's capability to operate at rated ranges, capacities, intensities, such as: revolutions per minute, pounds per square inch, temperature range, pounds of steam per minute, etc.

Operation qualification protocol
An operation qualification protocol (OQ) contains the plan and details of procedures to verify specific dynamic attributes of a utility/system, or process equipment, throughout its operated range including worst case conditions. Operation qualification when executed is documented verification that the system or subsystem performs as intended throughout all anticipated operating ranges.

Operating range
A range of values for a given process parameter that lie at or below a specified maximum operating value and/or at or above a specified minimum operating value, and are specified on the production worksheet or the standard operating instruction.

Overkill sterilization process
A process which is sufficient to provide at least a 12 log reduction of micro-organisms having a minimum D-value of 1 minute.

Processes
Processes are those activities which are repeated frequently such as pH-adjustment, including the preparation of solutions which are used for adjusting the pH; cleaning in place (CIP), and the preparation of CIP solutions; the various piping adjustments required to direct the solutions, sanitizing/sterilizing in place (SIP) and supportive activities; any sterilization of product, component, garment, equipment, etc., and any electromechanical or computer assisted processes associated with them.

Process equipment
Process equipment means such items as scales, load cells, flow meters, reaction/process/storage vessels, centrifuges, filters, driers, packaging equipment including electromechanical or computer assisted instruments, controls, monitors, recorders, alarms, displays, interlocks, etc., which are used in the manufacture of pharmaceutical products.

Process parameters
Process parameters are the properties or features that can be assigned values that are used as control levels or operating limits. Process parameters assure that the product meets the desired specifications and quality. Examples might be: pressure at 5.2 psig, temperature at $37 \pm 0.5°C$, flow rate at $10 \pm 1.0GPM$, pH at 7.0 ± 0.2.

Process variable
Process variables are the properties or features of a process which are not controlled or which change in time or by demand; process variables do not change product specifications or quality.

Process validation
Establishing documented evidence which provides a high degree of assurance that a specific process will consistently produce a product meeting its predetermined specifications and quality attributes.

Process validation protocol
Process validation protocol (PV) is a documented plan and details of procedures to verify specific capabilities of a process equipment/system through the use of simulation materials such as, for example, the use of a nutrient broth in the validation of an aseptic filling process, or the use of a placebo formulation in a freeze-drying process. However, the product as the material can be used to validate the process.

Product validation
A product is considered validated in the USA after completion of three successive successful full lot size attempts. These validation lots are saleable.

Prospective validation
Validation conducted prior to the distribution of either a new product, or product made under a revised manufacturing process, where the revisions may have affected the product's characteristics, to ensure that the finished product meets all release requirements for functionality and safety.

Protocol
A protocol is defined in this book as a written plan stating how validation will be conducted.

Quality assurance
The activity of providing evidence that all the information necessary to determine that the product is fit for the intended use is gathered, evaluated, and approved. The quality assurance department executes this function.

Quality attributes
Quality attributes refer to those properties of the product of a utility/system, such as resistivity of a water solvent, particulate matter, microbial and endotoxin limits of water for injection.

Quality control
The activity of measuring process and product parameters for comparison with specified standards to assure that they are within predetermined limits and, therefore, the product is acceptable for use. The quality assurance department executes this function.

Retrospective validation
Validation of a process for a product already in distribution based upon establishing documented evidence, through review/analysis of historical manufacturing and product testing data, to verify that a specific process can consistently produce, meeting its predetermined specifications and quality attributes.

In some cases a product may have been on the market without sufficient premarket process validation.

Retrospective validation can also be useful to augment initial premarket prospective validation for new products or changed processes.

Revalidation
Repetition of the validation process or a specific portion of it.

Specifications
Document which defines what something is by quantitatively measured values. Specifications are used to define raw materials, in-process materials, products, equipment and systems.

Standard operating procedure (SOP)
Written procedures followed by trained operators to perform a step, operation, process, compounding or other discrete function in the manufacture or production of a bulk pharmaceutical chemical, biologic, drug, or drug product.

State of control
A condition in which all process parameters that can affect performance remain within such ranges that the process performs consistently and as intended.

Static attributes
Static attributes may include conformance to a concept, design, code, practice, craftsmanship, material/finish/installation specifications, and absence of unauthorized modifications.

Sterilizing filter
A filter which, when challenged with the microorganism *Pseudomonas diminuta*, at a minimum concentration of 10 organisms per cm^2 of filter surface, will produce a sterile effluent.

Utilities/systems
Utilities/systems include such things as heating, ventilation, and air conditioning (HVAC) systems, process water, product water (purified water, USP; water for injection), clean steam, process air, vacuum, gases, etc., and electromechanical or computer assisted instruments, controls, monitors, recorders, alarms, displays, interlocks, etc., which are associated with them.

Validation
Establishing documented evidence which provides a high degree of assurance that a specific process will consistently produce a product meeting its predetermined specifications and quality.

Validation programme
The collective activities that are related to validation.

Validation protocols
Validation protocols are written plans stating how validation will be conducted, including test parameters, product characteristics, production equipment, and decision points on what constitutes acceptance results. This definition is provided by the US FDA. A maximum of four protocols are possible. They are protocols for installation qualification, operation qualification, process validation, and product validation. When the protocols have been executed it is intended to produce documented evidence that the system has been validated.

Validation scope
The scope answers the questions, what are to be validated? In the instance of the manufacturing plant, this would include the elements which impact critically on the quality of the product.

The elements which require validation are facilities, utilities/systems, process equipment, process and product.

314

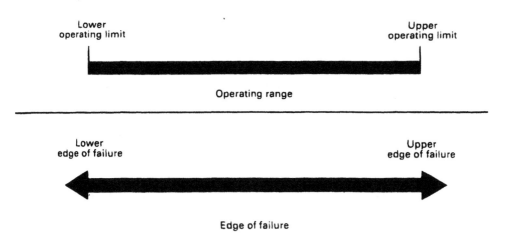

Figure 15.3 Worst case conditions

Worst case

A set of conditions encompassing upper and lower processing limits and circumstances, including those within standard operating procedures, which pose the greatest chance of process or product failure when compared to ideal conditions (see Figure 15.3). Such conditions do not necessarily induce product or process failure.

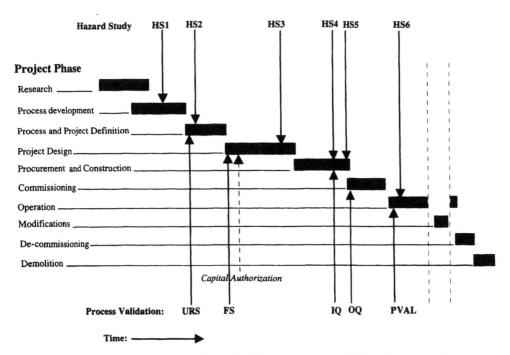

Figure 15.4 The six-stage hazard study methodology and process validation for a typical pharmaceutical project

Hazard Study and Validation Synergy

John Gillett (1994), in a series of papers, has provided some suggestions for improving the synergy between hazard study and various validation methods. The objective is to reduce the time taken in validation documentation preparation and studies by extraction, much of the information required from the Hazop process.

Hazard study and process validation have had different histories during their evolution. During the last decades, however, the two methodologies have drawn closer together in the pharmaceutical industry so that they overlap in several areas. Figure 15.4 shows these areas of overlap diagrammatically. The diagram represents a six-stage hazard study applied to a typical pharmaceutical project life-cycle with the associated validation activities included.

The six-stage hazard study consists of Hazard Study 1 (HS1) to get the facts, Hazard Study 2 (HS2) to identify significant hazards, Hazard Study 3 (HS3) to perform a hazard and operability study of the final design, Hazard Study 4 (HS4) and Hazard Study 5 (HS5) to check that the hazards identified have been controlled to acceptable standards, and Hazard Study 6 (HS6) to review the project and lessons learned.

Process validation starts with the preparation of a user requirements specification (URS) followed by a functional specification (FS) for engineering design and procurement. Installation qualification (IQ) and operation qualification (OQ) are performed to prove that the URS and FS have been met prior to the final process qualification or process validation. The reader is recommended to read the full paper (Gillet, 1994).

Preliminary Drawings/Facility Design

Here a general layout of the facility should be shown which depicts the people, material, and product flows respectively. This section of the master plan should describe how the material, people, and the product are manipulated and what constraints are applied to these flows to ensure compliance to cGMP. In particular, the handling of toxic material should be detailed. The HVAC and vacuum systems can be equipped with a HEPA filtration capability to remove toxic materials from the exhausted air and the rooms which are exposed to toxic materials maintained under negative pressure to further contain hazardous substances. Sluice/air shower areas are judiciously deployed for personnel to gown and degown to and from the critical areas. Areas throughout the facility can be designated red, yellow, or green depending on the nature of the activities taking place. Designations in certain areas fluctuate as the activities change. The equipment and process can also be selected and designed to provide an enclosed system to further contain toxic materials. Dedicated ventilation systems for hazardous materials can also be employed in the facility. Equipment must be thoroughly cleaned between campaigns according to written standard operating procedures (SOPs), and dedicated equipment should be used where cleaning is difficult.

Raw Materials Qualification

This section of the master plan is concerned with the raw materials and their testing before acceptance for use in the process. Chemical raw materials, filters, and packaging for products in the facility must be qualified before use. In all cases, qualification consists of the establishment of specifications for all materials and their subsequent testing. A minimum of three batches or lots per vendor are required to consecutively and successfully pass testing before a material can be considered qualified and be used for production. Subsequent lots are all tested against the established specifications before use.

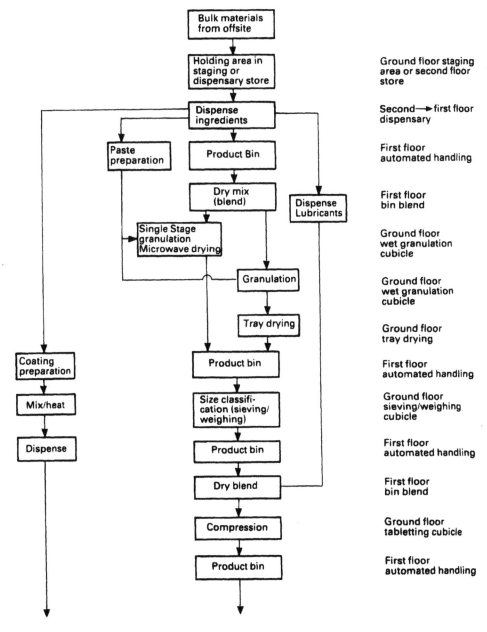

Figure 15.5 Process plan: preparation and packaging of sugar coated tablets, using wet granulated methods

317

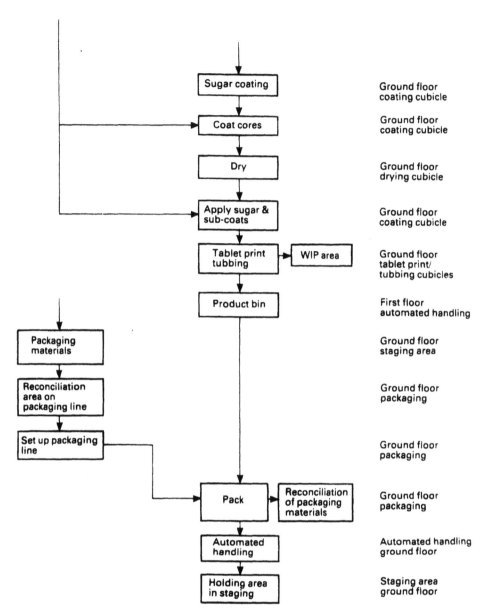

Figure 15.5 (continued)

The specifications for raw materials used in the preparation of solid oral dosage forms consist of a number of properties such as bioburden, particle size, bulk density, per cent moisture, and infrared spectra. This is not an exhaustive list, and it depends on the raw material, and the product in which it is used.

Certificates of analysis must be obtained from each Vendor on every lot of raw material. An audit of each vendor's manufacturing operation and process must also be conducted. All chemical raw materials are stored under controlled conditions to protect materials from excesses in temperature, humidity, etc., which are likely to affect stability.

Specifications for all compendial materials (USP XXII, etc.) are based upon the associated compendial monograph. The specifications, test procedures, handling, sampling storage, control, and release of all chemical raw materials must be covered by written standard operating procedures (SOPs). A written procedure covering the retesting of materials after storage for prescribed periods would also be prepared. Samples of all raw materials are retained in sufficient quantity to permit a minimum of two complete retests to ensure compliance with CFR 21.

Where stratification of raw materials is a concern, samples are taken from the top, middle, and bottom of the respective containers. Vendors are required to standardize on the type of container/closure/liner employed, with no variation permitted without qualification.

Where chemicals, raw materials, and compounds have toxic properties, special precautions must be taken in their handling, and laboratories must be suitably equipped.

Process Description

In this section there should be a number of paragraphs which highlight the critical process stages in the production of a particular dosage form and its subsequent packaging operation. A block flow diagram can be used to illustrate these stages. Figure 15.5 illustrates this concept.

Rooms and Room Classifications

Table 15.2 is an example of the type of information required in the master plan. It lists the rooms, by number and name, and a series of headings is provided under which the appropriate HVAC design parameter should be inserted for those areas in the facility subject to cGMP control considerations.

Description of Utilities

This section should include a brief outline of all the process utility systems that will be subject to validation. An indication of the depth of content is given here to provide the creation of the Master Plan with a starting point.

Purified Water System

The purified water system is used to produce water meeting or exceeding the specifications for purified water in *United States Pharmacopoeia* XXII, page 1457.

Such a system typically consists of generating equipment, storage facility, and filter units. A separate flow and return line supplies a distribution loop to points of use with a hygienic pump and UV unit on the supply line. The return line will also use a UV unit. All lines will be self-draining and designed free of 'dead legs'.

The generating system can consist of a 20 μm prefilter, an activated carbon filter, and a UV unit, 2 twin bed anion/cation resin exchanges, a 1 μm and 0.4 μm filter

Table 15.2 HVAC design basis

Room No.	Room name	Class	Summer DB	RH	Winter DB	RH	Filter	A/C per hour	Pressure	HVAC unit	Remarks
001/2	Utilities room										
003	Lift motor room										
004	Main stairs										
006	Emergency stairs										
007	Corridor/lobbies										
008	Centrifuge room										
009	Sluice & shower										
010/11	Mill wash										
012	Dryer/discharge										
013	Phosgene										
014	Tablet compressing										
015	Centrifuge										
016	Camfil filter										
101	Control room										
102	Office										
103	WC/toilet										

A/C = Air changes; DB = Dry bulb temperature °C; RH = Relative humidity %

unit, a UV unit, and a stainless steel storage tank with a design capacity sufficient to meet production requirements and provide spare capacity in case of faults.

Potable water will initially pass through a multi-media filter to remove large particulates, and then to a carbon bed to remove chlorine and organics. The multi-media filter and carbon bed will have provision for periodic back-flushing and steam sanitization respectively. If a reverse osmosis unit is used then a brief description of its *modus operandi* should be included, e.g. the reverse osmosis unit will remove over 95% of ionic contaminants and over 99% of pyrogens, colloidal silica and organics greater than 200 in molecular weight. This treated water is supplied to a storage tank which, in turn, feeds a mixed bed ion exchange polishing unit, a UV unit, submicrometre filters, and a return to the storage tank to complete the water treatment system distribution loop. Effluent from the submicron filters will be tapped to supply points-of-use.

Heating, Ventilating and Air Conditioning

The heating, ventilating, and air conditioning (HVAC) system is used to supply environmental air to manufacturing areas and to maintain appropriate pressure differentials. Process areas generally are pressurized negatively with respect to adjoining non-process areas in order to ensure containment. The HVAC system consists of air handlers, ductwork, fans, filters, heating and cooling coils, gauges, and controls. HEPA filtration units will be used in critical process areas, and airlocks are employed to separate areas of differing classifications. Cleaner, more critical, areas will be pressurized positively with respect to less clean areas. HVAC support to critical aseptic processing areas is capable of supplying air meeting Class 100 specifications contained in Federal Standard 209E. Immediate support areas will be Class 10 000.

The HVAC system is designed to control temperature and humidity for comfort and as appropriate for particular processes and products.

Air changes and a percentage recirculation are as appropriate for the area classification and processes. Critical Class 100 areas have an alarm system to detect malfunctions. Typically, a number of air-handling units (AHUs) supply air to all manufacturing storage, laboratories, and office areas. The AHU consists of an input fan, a filter system depending on the quality of air required, an extract fan, a duct distribution system, and control instrumentation designed to operate the system in the most efficient and cost effective manner.

Generally, for a process plant producing a number of different products, recirculation is not permitted owing to cross-contamination concerns.

Plant Steam System

The plant steam system is used to produce both high and low pressure steam for the facility. The primary function of the high pressure steam is to feed the clean steam generator system. Low pressure steam is used for the HVAC system, and to produce hot water for injection (WFI) tank and filter housing, glasswasher stations, and clean in place (CIP) system. The steam distribution system consists of insulated carbon steel pipes for feeding preheated water to both high and low pressure boilers. Appropriate valves and regulators are installed for proper maintenance and safety reasons as needed. Major components of the plant steam generator system are as follows.

The boiler feedwater package includes a storage tank and is fed from a water softener, condensate return from the clean steam generator system, and condensate from the low pressure boiler distribution. The package is equipped with a drain line.

The high pressure boiler package is capable of producing sufficient steam. (The design capacity should be given for both high pressure and low pressure boiler packages.) The chemical feed system feeds the necessary chemicals to the boiler.

The low pressure boiler package is fed with pre-heated water from the boiler feedwater system. This system generates steam to be used in the hot WFI tank, jackets and filter housings, HVAC system, CIP system, and in the processing areas where steam is needed.

The chemical feed system feeds the appropriate chemicals to the high and low pressure boilers. It consists of three identical systems each equipped with agitator, metering pump, and suction piping and a storage tank.

Provisions are made for blowdown of the boilers through a 'boiler blowdown' system which is equipped with an appropriately sized drain line.

At each point of use steam lines will come off the top of the header and steam traps will be judiciously placed where possible to eliminate condensate.

Chilled Water/Glycol Process

The chilled water generation system in the facility provides chilled water at 4°C to the chilled water distribution system. The distribution system, through insulated carbon steel piping, provides the chilled water to points-of-use throughout the facility, but not the HVAC system, and returns the water to the chilled water generation system for recirculation.

The chilled water generation system consists of two separate loops that exchange heat through a shell and tube heat exchanger. The capacity of the heat exchanger is sufficient for the system, and it receives fresh and returned chilled water at the tube side heat exchanger, using two motor driven centrifugal pumps.

40% propylene glycol, which has been cooled to 4°C in the process chiller, is pumped through the shell side of the heat exchanger via two motor driven centrifugal pumps and is used to cool the chilled water loop. The chilled water loop that receives the returned chilled water from the plant and made-up water from the water softener system in a carbon steel expansion tank, circulates this water through the entire chilled water generation, and the distribution system is equipped with safety release valves, level sensor in the tanks, temperature control probes, and appropriate valves and temperature or pressure indicators.

Nitrogen System

The nitrogen system will be used to produce pharmaceutical grade nitrogen suitable for product contact and meeting or exceeding the specifications presented in *United States Pharmacopeia* XXII p. 1952.

Table 15.3 Equipment requiring validation

Equipment use title	Product contact (surface construction material)	Non-contact surface (construction material min. standard)	Room no.
Mixing vessel	316L stainless steel	306 stainless steel, aluminium alloy	415
Pump	PTFE	306 stainless steel body, carbon steel base	318
Mill	316 stainless steel hammers mill chamber entry and discharge ports	306 stainless steel supports	205

The system will consist of a liquid nitrogen tank and distribution network supplying the points-of-use. Piping will be sloped-to-drain and constructed of 316L stainless steel.

Vacuum Cleaning System

The vacuum cleaning system will be a fixed piped system with the vacuum blower in the plant room, two vacuum filters, and six vacuum points located throughout the building. The vacuum blower will be controlled by six control stations adjacent to each vacuum point. Each control station will comprise an on/off switch and a 'run' light. The start/stop switches will be connected in parallel and the autostation will accept one signal. The 'run' lights will also be parallel, and this outstation will provide a DC supply to the light.

Low Temperature Coolant

The low temperature coolant (LTC) is a 50% solution of methanol and water. The LTC chillers are non-hermetic in construction, employing an evaporator, liquid receiver, and evaporative condenser. The LTC storage tank capacity is 18 000 L (this value will depend on the requirements of the system) vertical cylindrical vessel divided by a central perforated baffle, and it is fully drainable. The existing LTC chillers and circulation pumps will be incorporated into the new system which will include the installation of two new chillers. The LTC pipework is constructed of carbon steel and is protected from corrosion by dosing with chemicals. The LTC system will service the reactors and the thermal liquid heaters, condenser, precondenser, solvent charge vessel, and solvent condenser. This low temperature system is designed to flow at $-30°C$ with a return temperature of $-25°C$.

Table 15.4   Protocol outline: Purified Water USP

Installation qualification	Operation qualification	Process/product validation
• Approval page	• Approval page	• Approval page
• System description	• System description	• System description
• Statement of purpose	• Statement of purpose	• Statement of purpose
• Inspection checklist	• Divert-to-drain system	• Testing procedures
• Installation checklist	• Flow	chemical & physical
• Drawings	• Operation and capacity	constituents
• Materials of construction	• Alarms	• Testing procedures
• Weld reports	• Water softener operation	Microbiological
• Hydrostatic test reports	• Reverse osmosis operation	constituents
• Cleaning and passivation reports	• Ultraviolet sterilizer operation	• Sampling plan
• Manufacturer's certification	• Ultrafilter operation	• System sanitization
• Calibration review	• Filter integrity	• Carbon filter sanitization
• Standard operating procedures	• Miscellaneous testing	• Chlorine removal
• Training	• Acceptance criteria summary	• Flushing
• Supporting utilities	• Summary, analysis & certification	• Regeneration, back-washing & silica removal
• Expendables consumables	• Appendix reference Documents	• DI bed sanitization
• Space parts list		• Neutralization
• Punchlists		• Reverse osmosis effectiveness
• Summary, analysis & certification		• Ultraviolet disinfection effectiveness
• Appendix reference Documents		• Ultrafilter effectiveness
• Manuals		• Ultrafilter extractables
• Purchase orders		• Intensive monitoring
• Specifications		• Acceptance criteria
		• Summary, analysis, & certification
		• Appendix reference documents

Description of Process Equipment

A table should list all process equipment that will require validation. An example is given in Table 15.3.

The depth of information supplied so far in this chapter is an indication of what is required for the remainder of the headings shown in the introduction, e.g. training, SOP development, analytical support, calibration, protocol development, and the equipment history file. An example of the requirement for the protocol development is shown here for purified water USP and the HVAC systems.

Table 15.5   Protocol outline: HVAC

Installation qualification	Operation qualification	Process/product validation

• Approval page
• System description
• Statement of purpose
• Inspection checklist
• Installation checklist
• Drawings
• Materials of construction
• Cleaning reports
• Leak testing reports
• Air balancing reports
• Manufacturer's certification
• Calibration review
• Standard operating procedures
• Training
• Supporting utilities
• Expendables consumables
• Spare parts test
• Punchlists
• Summary, analysis & certification
• Appendix reference Documents
  • Manuals
  • Purchase orders
  • Specifications
	• Approval page
	• System description
	• Statement of purpose
	• Testing plan
	• Air handling units
	• Temperature and relative humidity
	• Particulate testing
	• Air velocity uniformity, and exchanges
	• Enclosure induction leak test
	• HEPA filter integrity test
	• Air flow patterns
	• Viable particle count
	• Differential room pressure
	• Recovery test
	• Alarms
	• Miscellaneous
	• Acceptance criteria summary
	• Summary, analysis, & certification
	• Appendix reference documents
		N/A

The stages shown in Tables 15.4 and 15.5 are required to validate any facility, process utility system equipment, and process.

Summary

- Prepare a master validation plan
- Provide complete protocols
- Develop the required SOP
- Provide training support in validation activities
- Prepare schedules of activities
- Administer the physical validation

Table 15.6 How validation interacts with project progress

Provider	Stages of the project Inception cost study	Design (DQ) document	Detailed design	Tender	Construction	Start up
Site design team	*Project team* Initial cost plan Local construction knowledge *Designers* Define the scope Agree strategy Liaise with client Preliminary programme Prepare preliminary layouts Process advice Structural, M & E strategy	*Project team* Liaise with client Local knowledge Cost information Local approvals Local standards Translations Structural information *Designers* Prepare room data sheets Confirm local regs Confirm building standards and techniques Review equipment suppliers Liaise with client Process advice surveys Confirm utilities	*Project team* Liaise with client Programming Work package identification Project control procedures Cost control *Designers* Liaise with client Gain client approvals Check drawings Discuss proposals with Local Authorities	*Project team* Liaise with client Contractor selection Translation of tender documents Tender preparation Cost control Contractor selection Translation of tender documents Review tender returns *Designers* Liaise with client Issue drawings Issue specifications Check tenders for design input	*Project team* Liaise with client Programming Contractor-work package control Control procedures Cost control Change order procedures Safety Construction valuations Final accounts *Designers* IQ/OQ control documents Building handover Equipment interface engineering Commissioning programming Quality inspections Site support	*Project team* Liaise with client Programming Control procedures Production moves Cost control Training Production start up
Office Design based	*Specialist design team* Finalize the cost report Start validation master plan (VMP) Area review Budget	*Specialist design team* Develop the brief Establish the cost plan Prepare layouts Agree quality and standards Complete DQ document Equipment information Room data sheets Corporate standards (USER) Client brief Client Sign Off	*Specialist design team* Prepare working drawings for tendering Prepare specifications Co-ordinate the design with process equipment Specialist kit information Layout development Special tender conditions Special requirements	*Specialist design team* Issue drawings and specifications Review tender returns	*Specialist design team* Site support Quality inspections Commissioning/validation support As built drawings Support Department move dates for programming, witnessing service sign off, etc.	Department move logistics GMP training requirements
Validation trail		DQ document			IQ/OQ documentation	PQ documentation

- Review and evaluate all data.
- Assemble the completed validation packages
- Complete the final certification of the facility, utilities, systems and processes.

Finally Table 15.6 shows how all the stages of a project interact with the validation process.

References

Good Automated Manufacturing Practice Document. Electronic version available from ISPE Den Haag, Netherlands.

D. W. MENDENHALL, Cleaning validation, *Drug Dev. Ind. Pharm.* 15 (13), 2105–2114 (1989).

K. M. JENKINS and A. J. VANDERWIELEN, Cleaning validation: an overall perspective, *Pharm. Technol.* 18 (4), 60–73 (1994).

G. L. FOURMAN and M. V. MULLEN Determining cleaning validation acceptance limits for pharmaceutical manufacturing operations, *Pharm. Technol.* 17 (4), 5406 (1993).

FDA Mid-Atlantic Region Inspection Guide on Cleaning Validation, revised 7 May 1993.

T. C. HWANT *et al.*, 'Investigational study for the potential factors on the cleaning results of pharmaceutical equipment,' poster presented at AAPS Midwest Regional Annual Meeting, Chicago, USA (1995).

PMA Cleaning Validation Survey Results, 1992.

R. C. HWANT *et al.*, Process design and data. Analysis for cleaning validation, *Pharm. Technol* 9 (2), 21–25 (1997).

M. C. OLVER 'Cleaning validation – a hands-on approach of validating a manual cleaning procedure', Computer and Process Validation '97, Manchester, 15–16 April 1997.

J. E. GILLET The six-stage Hazard Study Procedures and Process Validation, *Pharm. Technol.* 6 (9), 56–66 (1994).

Index

For Product Safety Concerns and Information please contact our EU
representative GPSR@taylorandfrancis.com
Taylor & Francis Verlag GmbH, Kaufingerstraße 24, 80331 München, Germany

www.ingramcontent.com/pod-product-compliance
Ingram Content Group UK Ltd.
Pitfield, Milton Keynes, MK11 3LW, UK
UKHW011456240425
457818UK00021B/855